FORMS/FORMAT 2010

Eckehard Schnieder · Géza Tarnai
Editors

FORMS/FORMAT 2010

Formal Methods for Automation and Safety in Railway and Automotive Systems

Springer

Editors

Eckehard Schnieder
Technische Universität Braunschweig
Institute for Traffic Safety
and Automation Engineering
Langer Kamp 8
D-38106 Braunschweig
Germany
e.schnieder@tu-braunschweig.de

Géza Tarnai
Budapest University of Technology
and Economics
Department of Control
and Transport Automation
Bertalan L. u. 2.
H-1111 Budapest
Hungary
tarnai.geza@mail.bme.hu

ISBN 978-3-642-14260-4 e-ISBN 978-3-642-14261-1
DOI 10.1007/978-3-642-14261-1
Springer Heidelberg Dordrecht London New York

Library of Congress Control Number: 2011921311

Cover design: WMXDesign GmbH, Heidelberg

Printed on acid-free paper

Springer is part of Springer Science+Business Media (www.springer.com)

Preface

Coping with the complexity of advanced automation- and safety systems both in railway and automotive applications will be more and more dominated by the use of formal means of description, formal methods, and tools. Altogether named Formal Techniques they provide next to the correctness and integrity checkups – especially in safety relevant systems – the possibility to prove the syntactic and semantic specification of the system as well as to simulate the system operation.

Formal methods – a comprehensive form for means of description and adjacent methodological concepts – gained by advanced and more and more professional tools supported by powerful computer technology emerge currently and find their benefits to lots of applications. Primarily, their promising power of clear description and symbolic patterns for modelling the real world including technological devices and human operators offers the chance for engineers to build up systems which can be designed in a correct way. With the quality of mathematical proofs they provide guaranteed conditions of dependability, the comprehensive term for availability, reliability, maintainability and safety and furthermore security, short RAMSS. In transportation, a high demand for RAMSS exists, especially under new European directives, regulation authorities, and standards on the one hand as well as expectations from the users' and operators' side on the other.

Requirements of the recently updated legal framework expressed by EU-Guidelines, IEC- and CENELEC-standards and establishing standards for automotive software which are based on formal techniques, particularly with regard to the handling of safety analysis, are to be treated in FORMS/FORMAT 2010. The main focus lies on topics facing formal techniques for railway applications and intelligent transportation systems as well as for automotive applications. Gained findings, experiences and also difficulties associated with the handling of the subject matter are to be shown.

Hence the meanwhile 8th Symposium of FORMS / FORMAT and its subtitle "Formal Methods for Automation and Safety in Railway and Automotive Systems" fully cover the broad joint approach for this challenging topic. Since transportation in its whole can profit from this theoretical approach for formal methods, the scope inside transportation has expanded for railway and road transportation, mainly tackled jointly by methodological approaches.

Supervised by an internationally highly ranked experts' program committee from America, Asia, and Europe some twenty contributions have been selected very critically for oral presentation and to be published in the proceedings of the 2010 symposium. The symposium will be framed by invited contributions by internationally leading experts from operators, assessors, and science.

The first part of the program starts with contributions about three different aspects of RAMSS. The first session covers both safety and security and their policy for application in the transportation domain. It is followed by the increasing influence of maintenance to operations and ends with its methods for evaluation and analysis of essential functions of railway operations systems as well as its infrastructure and vehicles.

The second part of the program covers general aspects. Beginning with remarks on legal framework and risk metrics, it is followed by methods for the development and simulation in the automotive domain. Theoretical contributions about the verification of programmable logic controllers (PLC) which become more and more attractive for the control in transportation together with tool chains for testing and development conclude the program.

The current proceedings include the papers of these different sessions of the Symposium FORMS/FORMAT 2010, which present novel research and practical results that have been reached since the previous symposium.

We would like to acknowledge the contribution of every attendant and the support of the program committee, and we hope for a prospering future of our common activity and also to widen the sphere of users again. We are convinced that our symposium will provide an invisible but nevertheless important contribution to safe transportation.

The editors thank all authors for their support, especially Geltmar von Buxhoeveden for his careful preparation of the symposium, and Springer Verlag for publishing the proceedings.

December 2010

Eckehard Schnieder, Géza Tarnai
Program Chairs
FORMS/FORMAT 2010

Conference Organization

Programme Chairs

Eckehard Schnieder, Géza Tarnai

Programme Committee

Marc Antoni	SNCF (F)
Joachim Axmann	Volkswagen AG (D)
Jens Braband	Siemens AG (D)
Henning Butz	Airbus Deutschland (D)
Werner Damm	Carl von Ossietzky Universität Oldenburg (D)
El-Miloudi El-Koursi	INRETS Lille (F)
Alessandro Fantechi	Università degli Studi di Firenze (I)
Martin Fränzle	Carl von Ossietzky Universität Oldenburg (D)
Peter Göhner	Universität Stuttgart (D)
Shigeto Hiraguri	RTRI - Railway Technical Research Institute (J)
Yuji Hirao	Nagaoka University of Technology (J)
Aleš Janota	University of Žilina (SK)
Karsten Lemmer	DLR Braunschweig (D)
István Majzik	Budapest University (HU)
Bin Ning	Beijing Jiaotong University (CN)
Jörn Pachl	Technische Universität Braunschweig (D)
Markos Papageorgiou	Technical University of Crete (GR)
Stefano Ricci	Università di Roma (I)
Bastian Schlich	RWTH Aachen (D)
Holger Schlingloff	Humboldt-Universität Berlin (D)
Roman Slovák	BAV - Bundesamt für Verkehr (CH)
Olaf Stursberg	Universität Kassel (D)
Aníbal Zanini	Universidad de Buenos Aires (AR)

Local Organization

Geltmar von Buxhoeveden, Christian Cholewa, Güler Gülec, Sylvia Glowania, Christine Jendritzka, Sofia Mouratidis, Sarah-Romina Pesenecker, Felix Reinbold, Arno G. Schielke, Sven Schulze, Nadine Schwarz, Andreas Siepmann, Akbar Shah, Regine Stegemann, Kevin Wieloch

External Reviewers

Lars Ebrecht, Matthias Grimm, Malte Hammerl, Christian Herde, Katrin Lüddecke, Michael Meyer zu Hörste, Markus Pelz

Table of Contents

Invited Lectures

Achieving Europe-wide safety through technical harmonization

Ralf Schweinsberg

Vicepresident Federal Railway Authority,
Eisenbahn-Bundesamt [EBA], Heinemannstraße 6, D 53175 Bonn
poststelle@eba.bund.de

Abstract. Technical and operational harmonization is beyond liberalization one of the key elements of the EU transport policy for the railway market. The main legal basis are the European Interoperability Directives and the Railway Safety Directive which regulate the certification and authorization processes for Railway Systems.
Mandatory Technical Specification for Interoperability that are elaborated by working groups of the European Railway Agency (ERA) define for the different subsystems (Infrastructure, Rolling Stock etc.) harmonized criteria or refer to European Standards, which become hereby also binding. The system of TSI follows a modular approach. A new locomotive has for example to fulfilll the relevant requirements of the TSI Rolling Stock, Control Command and Signalling, Persons with reduced mobility, Safety in Railway Tunnels, Operation and Noise. Future innovations should not be blocked by TSI. Due to this only those requirements should be regulated in TSI that are absolutely necessary for interoperability. The harmonization by voluntary European Standards which are listed in a TSI Application Guide support mutual acceptance in the same way but has the advantage of more flexibility in case of new technical solutions and complete the picture of necessary European harmonization.
The requirements in the TSI are checked by so called Notified Bodies. The result has to be accepted in the whole European Union and can without good reasons not be questioned on national level. This means that here we have a "European level of safety". But as long as not all requirements are harmonized and the interface to the national infrastructures need to be considered, an additional authorization for placing into service by the National Safety Authority of the Member States, in which the subsystem is placed into service, is necessary. This authorization include additional national requirements that are not harmonized and the check of the coherency between the different subsystems, i. E. between a Rolling Stock and the Infrastructure. To avoid discrimination, the Member States are obliged to notify these National Safety Rules to the European commission in order to be published in the Internet. The future tool is the European Database NOTIF for the notification of national railway safety and technical rules. In this respect NOTIF is, as an intermediate step, the basis for ensuring transparency regarding those regulations that are still not harmonized.

E. Schnieder, G. Tarnai (eds.), *FORMS/FORMAT 2010*,
DOI 10.1007/978-3-642-14261-1_1,

Up to now the TSI are related to the Trans European Network. With the new Interoperability Directive it is foreseen to widen the scope. Due to this it has to be decided whether new systems, i. E. new category of Rolling Stock and Lines need to be considered.
Safety can only be achieved, if all relevant risks can be handled by specific measures and as long as we have no common European Railway Infrastructure the Network specific aspects play an important role.
Cross Acceptance of Authorizations that are offered by the National Safety Authorities is one major element of the new "Interoperability Directive" of the European Union. Those requirements, that are not yet harmonized in TSI have to be classified against the national requirements of other Member States in A (cross - accepted), B and C (not cross accepted) - parameters. The result need to be fixed in a cross reference document which includes all relevant parameters as a basis for future international Rolling Stock projects. It is possible that this process in the future lead to harmonized requirements on a European level but it is nevertheless an important intermediate step because it leads to "harmonization" on bi- and multilateral levels.
Safety includes technical and process - oriented aspects. Beyond the authorization for placing into service by the responsible National Safety Authority, the Railway Undertakings (RU) and Infrastructure Manager (IM) are obliged by the European Safety Directive to implement a Safety Management System (SMS) based on common requirements, which include, among others, processes with which the RU and IM ensure that they control all risks of railway operation, including purchase of new material. In order to fulfill this RU and IM have for example for the concrete network access to check, that the specific Rolling Stock is compatible with the Infrastructure on which it is foreseen to be operated. Hereby the RU and IM identify for example the necessary Control Command and Signalling System. Only with this second step a safe integration can be ensured.
The European Common Safety Methods (CSM) on Risk evaluation and assessment, which have to be applied since 19.07.2010 according to a European Regulation enforce the sector to follow a structured process which include especially the identification and classification of possible hazards. The CSM also regulate the selection of one of the three known risk acceptance principles: Code of Practice, Reference System or Explicit Risk Analysis and based on the results the decision about the necessary safety measures to proof that all possible Hazards are controlled. The result has to be documented in a hazard log and assessed by an independent safety assessor. The CSM also foresees a permanent Hazard Management. With the implementation of this transparent and structured process cross acceptance of subsystems like for example Rolling Stock shall be supported. Everybody speaks "the same language"; a code of practice is for example identified in the same manner in all EU - Member States.

"Open Proof" for Railway Safety Software - A Potential Way-Out of Vendor Lock-in Advancing to Standardization, Transparency, and Software Security

Klaus-Rüdiger Hase

Deutsche Bahn AG, Richelstrasse 3, 80634 München, Germany
Klaus-Ruediger.Hase@DeutscheBahn.com

Abstract. "Open Proof" (OP) is a new approach for safety and security critical systems and a further development of the "Open Source Software" (OSS) movement, not just applying OSS licensing concepts to the final software products itself, but also to the entire life cycle and all software components involved, including tools, documentation for specification, verification, implementation, maintenance and in particular including safety case documents. A potential field of applying OP could be the European Train Control System (ETCS) the new signaling and Automatic Train Protection (ATP) system to replace some 20 national legacy signaling systems in all over the European Union. The OP approach might help manufacturers, train operators, infrastructure managers as well as safety authorities alike to eventually reach the ambitious goal of an unified fully interoperable and still affordable European Train Control and Signaling System, facilitating fast and reliable cross-border rail traffic at state of the art safety and security levels.

Keywords: ATC, ATP, Critical Software, ETCS, EUPL, Embedded Control, FLOSS, Open Proof, openETCS, Train Control, Standardization.

1 Introduction

The European Train Control System (ETCS, [1]) is intended to replace several national legacy signaling and train control systems all across Europe. The system consists of facilities in infrastructure and on-board units (OBU). Especially for the ETCS on-board equipment the degree of functional complexity to be implemented is expected to be significantly higher than in conventional systems. In terms of technology, this is mostly done by software in so-called embedded control system implementations. While electronic hardware is getting continuously cheaper, the high complexity of the safety critical software has caused significant cost increases for development, homologation and maintenance of this technology. This has raised questions for many railway operators with respect to the economy of ETCS in general.

E. Schnieder, G. Tarnai (eds.), *FORMS/FORMAT 2010*,
DOI 10.1007/978-3-642-14261-1_2, © Springer-Verlag Berlin Heidelberg 2011

The key element for improving that situation seems to be a greater degree of standardization in particular for the ETCS onboard equipment on various levels: Hardware, software, methods and tools. Standardization by applying open source licensing concepts will be the focus of this paper.

1.1 From National Diversity to European Standard

Looking back into history of signaling and automatic train protection (ATP) for mainline railways systems, in the past 40 years a major change in technology has taken place. In the early days of ATP almost all functions were implemented in hardware, starting with pure mechanical systems, advancing to electromechanical components and later on using solid state electronics, like gates, amplifiers, and other discrete components. Software was not an issue then. Beginning in the late 1970 years an increasing number of functions were shifted into software, executed by so called micro computers. Today the actual functions of such devices are almost entirely determined by software. The dramatic performance increase of microcomputers in the past 30 years on the one hand and rising demand for more functionality on the other hand, has caused a significant increase in complexity of those "embedded control systems" – how such devices are usually called. Furthermore, the development from purely "monitoring" safety protection systems, like the German INDUSI (later called PZB: "Punktförmige Zug-Beeinflussung") or similar systems in other European countries, which only monitor speed at certain critical points and eventually stop the train, if the driver has missed a halt signal or has exceeded a safe speed level, to a more or less (semi) Automatic Train Control (ATC) systems like the German continuous train control system, called LZB ("Linien-Zug-Beeinflussung"), which has increasingly shifted safety responsibility from the infrastructure into the vehicle control units. Displaying signal commands inside the vehicle on certain computer screens, so called "cab signaling", has resulted in greater independence from adverse weather conditions.

In all over Europe there are more than 20 different mostly not compatible signaling and train protection systems in use (figure 1). For internationally operating high speed passenger trains or cargo locomotives up to 7 different sets of equipment have been installed, just to operate in three or four countries. Since each of those systems have their own "antennas" to sense signals coming from the way-side and their own data processing units and display interfaces, space limitations are making it simply impossible to equip a locomotive for operation in all EU railway networks, not to mention prohibitive cost figures for such equipment. Furthermore, some of the systems are in use for more than 70 years and may not meet today's expected safety level. Some are reaching their useful end of life causing obsolescence problems.

For a unified European rail system it is very costly to maintain this diversity of signaling systems forever and therefore the European Commission has set new rules by so called "Technical Specifications for Interoperability" (TSI) with the goal to implement a unified "European Train Control System",

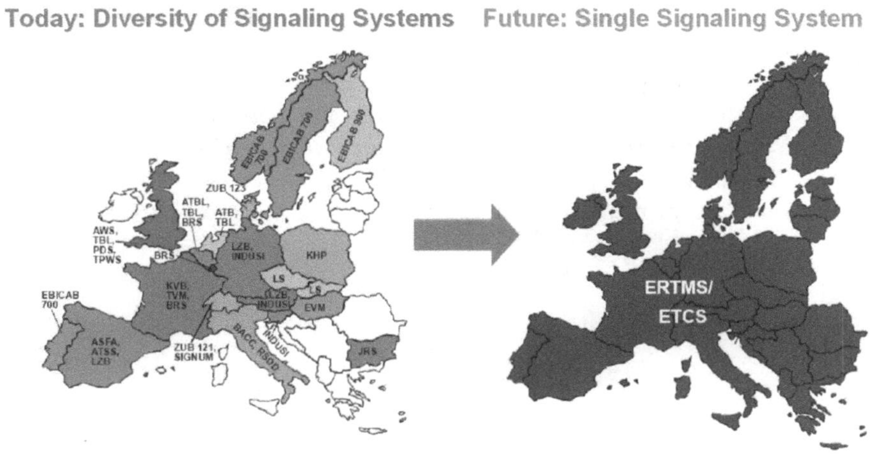

Fig. 1. Europe's challenge is to substitute more than 20 signaling and ATP systems by just one single system, ETCS, in order to provide border crossing interstate rail transit in all over the European Union.

which is part of the "European Rail Traffic Management System" (ERTMS), consisting of ETCS, GSM-R, a cab radio system based on the GSM public standard enhanced by certain rail specific extensions and the "European Traffic Management Layer" (ETML). Legacy ATP or so called "Class B" systems are supposed to be phased out within the next decades.

1.2 ETCS: A new Challenge for Europe's Railways

Before launching the ETCS program, national operational rules for the railway operation were very closely linked with the technical design of the signal and train protection systems. That is going to change radically with ETCS. One single technology has to serve several different sets of operational rules and even safety philosophies.

The experience of Deutsche Bahn AG after German reunification has made very clear that it will take several years or even decades to harmonize operational rules in all over Europe. Even under nearly ideal conditions (one language, one national safety board and even within one single organization) it was a slow and laborious process to convert different rules and regulations back into one set of unified operational rules. After 40-years of separation into two independent railway organizations (Deutsche Reichsbahn in the east and Deutsche Bundesbahn, west), it took almost 15 years for Deutsche Bahn AG to get back to one single unified signaling handbook for the entire infrastructure of what is today DB Netz AG.

Therefore, it seem unrealistic to assume that there will be one set of operational rules for all ETCS lines in all over Europe any time soon (Which does not mean that these efforts should not be started as soon as possible, but without raising far too high expectations about when this will be accomplished.). That means, in order to achieve interoperability by using a single technical solution: This new system has to cope with various operational regimes for the foreseeable future. Beside this, for more than a decade there will be hundreds of transition points between track sections equipped with ETCS and sections with one of several different legacy systems. This will cause an additional increase of functional complexity for onboard devices.

1.3 Technology is not the Limiting Factor

With state of the art microcomputer technology, from a technological point of view, this degree of complexity will most likely not cause any performance problems since the enormous increase in performance of microcomputer technology in recent years can provide more than sufficient computing power and storage capacity at an acceptable cost level; to master complex algorithms and a huge amount of data.

The real limiting factor here is the "human brain power". In the end it is the human mind, which has to specify these functions consistently and completely, then provide for correct designs, implement them and ultimately make sure that the system is fit for its purpose and can prove its safety and security. The tremendous increase in complexity, absorbing large numbers of engineering hours is one reason why we are observing cost figures for R&D, testing and homologation of the software components in various ETCS projects that have surpassed all other cost positions for hardware design, manufacturing and installation. This has caused a substantial cost increase for the new onboard equipment compared with legacy systems of similar functionality and performance.

Normally we would expect from any new technology a much better price to performance ratio than for the legacy technology to be replaced. Due to the fact, that this is obviously not the case for ETCS, makes it less attractive for those countries and infrastructure managers, who have already implemented a reliably performing and sufficiently safe signaling and train control system. In addition there is no improvement expected for ETCS with respect to performance and safety compared with service proven systems like LZB and PZB. In order to reach the goal of EU-wide interoperability soon, the EU Commission has implemented legal measures, regulating member state's policies for infrastructure financing and vehicle homologation. While in the long run, ETCS can lower the cost for infrastructure operators, especially for Level 2 implementations making conventional line signals obsolete, the burden of cost increase stays with the vehicle owners and operators.

Therefore it became an important issue for vehicle operators to identify potential cost drivers and options for cost reduction measures, so as not to endanger the well-intentioned general goal of unrestricted interoperability.

2 Software in ETCS Vehicle Equipment

As discussed above, state of the art technology requires for almost all safety critical as well as non-safety related functions to be implemented in software. The end-user will normally receive this software not as a separate package, but integrated in his embedded control device. Therefore software is usually only provided in a format directly executable by the built-in microprocessor, a patter of ones and zeros, but therefore not well suited for humans to understand the algorithm. The original source code, a comprehensible documentation format of this software, which is used to generate the executable code by a compiler and all needed software maintenance tools are usually not made available to the users. Manufacturers are doing this, because they believe that they can protect their high R&D investment this way.

2.1 Impact of "Closed Source" Software

However concealment of the software source code documentation has increasingly been considered as problematic, not only for economical reasons for the users, but more and more for safety and security reasons as well.

Economically it is unsatisfactory for the operators to remain completely dependent from the original equipment manufacturer (OEM), no matter whether software defects have to be fixed or functions to be adapted due to changing operational requirements. For all services linked to these embedded control systems there is no competitive market, since bundling of non-standard electronic hardware together with "closed source" or "proprietary" software makes it practically and legally impossible for third parties to provide such service. This keeps prices at high levels.

While malfunctions and vulnerability of software products, allowing malware ("malicious software": as there are viruses, trojans, worms etc.) to harm the system, can be considered as quality deficiencies, which can practically not be discovered in proprietary software by users or independent third parties, whereas the question of the "vendor lock-in" due to contractual restrictions and limiting license agreements is generally foreseeable, but due to generally accepted practices in this particular market, hardly to be influenced by individual customers (e.g. railway operators).

Especially security vulnerability of software must be considered as a specific characteristic of "proprietary" or "closed source" software. So-called "End User License Agreements" (EULA) do usually not allow end-users to analyze copy or redistribute the software freely and legally. Even analysis and improvement of the software for the user's own purposes is almost generally prohibited in most EULAs. While on the one hand customers who are playing by the rules are barred from analyzing and finding potential security gaps or hazardous software parts and there-fore not being able to contribute to software improvements, even not for obvious defects, however the same legal restrictions on the other hand do not prevent "bad guys" from disassembling (a method of reverse-engineering) and analyzing the code by using freely available tools, in order to search for security gaps and occasionally (or better: mostly) being successful in finding unauthorized access points or so-called "backdoors". Intentionally implemented backdoors by irregularly working programmers or just due to lax quality assurance enforcement or simply by mistake are causing serious threats in all software projects. In most cases intentionally implemented backdoors are hard to find with conventional review methods and testing procedures. In a typical proprietary R&D environment only limited resources are allocated in commercial projects for this type of "security checks" and therefore stay most likely undiscovered.

That backdoors cannot be considered as a minor issue, has been discussed in various papers [2], [3], [4], [5] and has already been identified as a serious threat by the EU Parliament, which has initiated resolution A5-0264/2001 in the aftermath of the "Echelon" interception scandal, resulting in following recommendations [6]:

"... **Measures to encourage self-protection by citizens and firms:**

29. *Urges the Commission and Member States to devise appropriate measures to promote ... and ... above all to support projects aimed at developing* ***user-friendly open-source encryption software;***
30. *Calls on the Commission and Member States to promote software projects whose* ***source text is made public (open-source software), as this is the only way of guaranteeing that no backdoors are built into programmes;***
31. *Calls on the Commission to lay down a* ***standard for the level of security*** *of e-mail software packages, placing those packages* ***whose source code has not been made public in the least reliable category;...***"

This resolution was mainly targeting electronic communication with private or business related content, which most likely will not hurt people or endanger their lives. However a recent attack by the so called "STUXNET" worm [7], a new type of highly sophisticated and extremely aggressive malware, which in particular was targeting industrial process control systems via its tools chain, even in safety critical applications (chemical and nuclear facilities). Systems, which are very similar in terms of architecture and software design standards with signaling and interlocking control systems. This incident has demonstrated that we have to consider such impact in railway control and command systems as well, commercially and technically.

2.2 Software Quality Issues in ETCS Projects

Despite a relatively short track record of ETCS in revenue service we had already received reports on accidents caused by software defects, like the well documented derailment of cargo train No. 43647 on 16 October 2007 at the Lötschberg base line in Switzerland [8].

German Railways has been spared so far from software errors with severe consequences, possibly due to a relatively conservative migration strategy.

During the past 40 years, software was only introduced very slowly in small incremental steps into safety-critical signaling and train protection systems and carefully monitored over years of operation, before rolled out in larger quantities. Software was more or less replacing hard-wired circuits with relatively low complexity based on well service-proven functional requirement specifications over a period of four decades.

With ETCS however, a relatively large step will be taken: Virtually all new vehicles have to be equipped with ETCS from 2015 on, enforced by European legal requirements, despite the fact that no long-term experience has been made.

The ongoing development of the functional ETCS specification as well as project-specific adaptations to national or line-specific conditions has resulted in numerous different versions of ETCS implementations not fully interoperable. Up to now, there is still no single ETCS onboard equipment on the market that could be used on all lines in Europe, which are said to be equipped with ETCS. That means that the big goal of unrestricted interoperability would have been missed, at least until 2010.

The next major release of the System Requirements Specification (SRS 3.0.0), also called "baseline 3", is expected to elimination those shortcomings.

Baseline 3 has another important feature: Other than all previous SRS versions, which have been published under the copyright of UNISIG, an association of major European signaling manufacturers, SRS 3.0.0 in the opposite has been published as an official document by the European Railway Agency (ERA) a governmental organization implemented by the European Commission. This gives the SRS a status of a "public domain" document. That means, everyone in Europe is legally entitled to use that information in order to build ETCS compliant equipment.

2.3 Quality Deficiencies in Software Products

Everyone who has ever used software products knows that almost all software has errors and no respectable software company claims, that their software is totally free of defects.

There are various opportunities to make mistakes during the life cycle of software production and maintenance: Starting with System Analysis, System Requirement Specification, Functional Requirement Specification, etc., down to the software code generation, integration, commissioning, operation and maintenance phases. A "NASA Study on Flight Software Complexity" [12] shows contribution to bug counts, which can be expected in different steps of software production (figure 2).

Figure 3 characterizes the actual situation in the European signaling industry with several equipment manufacturers working in parallel, using the same specification document in "natural language precision" giving room for interpretation, combined with different ways and traditions of making mistakes, resulting in a low degree of standardization, even for those components,

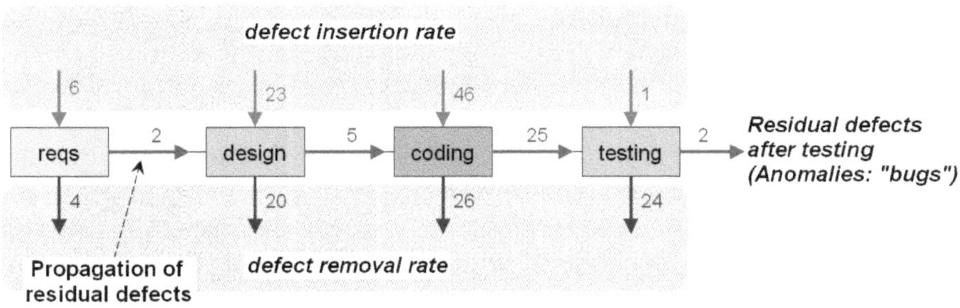

Fig. 2. Propagation of residual defects (bugs) as a result of defect insertion and defect removal rates during several stages of the software production process, according to a NASA research on high assurance software for flight control units for each 1000 lines of code (TLOC) [12].

which cannot be used for product differentiation (core functionality according to UNISIG subset 026 [1]). Since all or at least most of the documents are created by humans, there is always the "human factor" involved, causing ambiguities and therefore divergent results.

Herbert Klaeren refers to reports in his lecture [9], which have found an average of 25 errors per 1000 Lines Of programming Code (TLOC) for newly produced software. The book "Code Complete" by Steve McConnell has a brief section about errors to be expected. He basically says that there is a wide range [10]:

(a) *Industry Average: "about 15 - 50 errors per 1000 lines of delivered code." He further says this is usually representative of code that has some level of structured programming behind it, but probably includes a mix of coding techniques.*
(b) *Microsoft Applications: "about 10 - 20 defects per 1000 lines of code during in-house testing, and 0.5 defect per TLOC in released products [10]." He attributes this to a combination of code-reading techniques and independent testing.*
(c) *"Harlan Mills pioneered a so called 'clean room development', a technique that has been able to achieve rates as low as 3 defects per 1000 lines of code during in-house testing and 0.1 defect per 1000 lines of code in released product (Cobb and Mills 1990 [11]). A few projects - for example, the space-shuttle software - have achieved a level of 0 defects in 500,000 lines of code using a system of formal development methods, peer reviews, and statistical testing."*

However the U.S. space shuttle software program came at a cost level of about U.S. $ 1,000 per line of code (3 million LOC $\sim$ 3 billion U.S. $ [9], cost basis 1978), not typical for the railway sector, which is more in a range between 30 € per LOC for non-safety applications and up to 100 € for SIL 3-4 quality (SIL: Safety Integrity Level) levels.

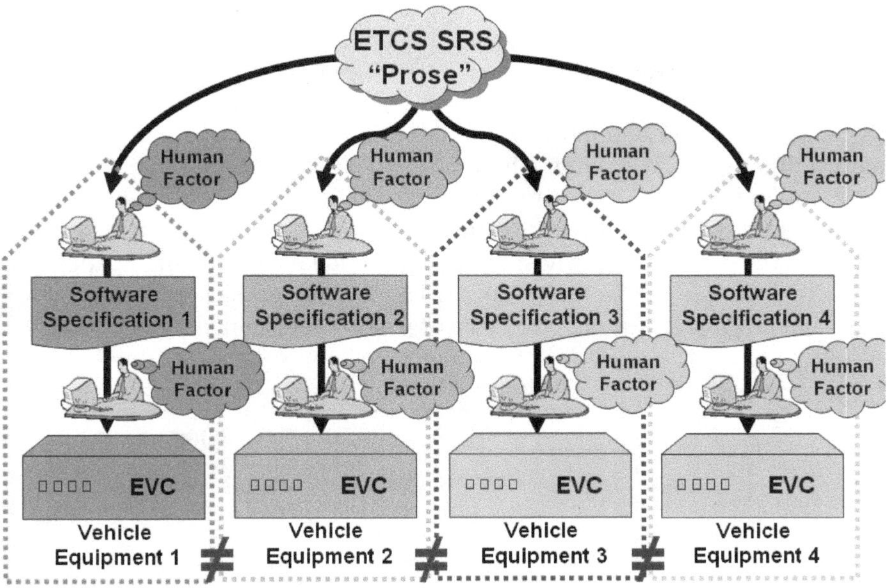

Fig. 3. Divergent interpretation of a common public domain ETCS System Requirement Specification (SRS) document, due to the "human factor" by all parties involved, causing different software solutions with deviant behavior of products from different manufacturers, which result in interoperability deficiencies and costly subsequent improvement activities.

2.4 Life Cycle of Complex Software

While on the one hand, electronic components are becoming increasingly powerful, yet lower in cost, on the other hand, cost levels of complex software products are increasingly rising not only because the amount of code lines need tremendous men power, but for those lines of code a poof of correctness, or also called "safety case", has to be delivered in order to reach approval for operation in revenue service. Some manufacturers have already reported software volumes of over 250 TLOC for the ETCS core functionality defined by ETCS SRS subset 026 [1].

It is very difficult to receive reliable statistics about errors on safety related soft-ware products, because almost all software manufacturers hide their source code using proprietary license agreements. However we can assume that software in other mission critical systems, like communication servers, may have the same characteristics with respect to "bug counts". One of those rare examples published, was taken from an AT&T communication software project, which from its size is in the same order of magnitude as today's ETCS onboard software packages (figure 4) [9],[13].

One particular characteristic in figure 4 is quite obvious: The size of the software is continuously growing from version to version, despite the fact that this software was always serving the same purpose. Starting with a relatively high bug count of almost 1000 bugs in less than 150 TLOC, the software "matures" after several release changes and is reaching a residual bug density, which is less than a tenth of the initial bug density. During its early life the absolute number of bugs is oscillating and stabilizes in its more mature life period. At a later phase of the life cycle the absolute number of bugs is slightly growing despite a decreasing bug density. The late life-cycle bug density is mainly determined by the effectiveness and quality measures taken on the one hand and the number of functional changes and adaptations built in since last release on the other hand. The actual number of bugs is often unknown and can only subsequently been determined.

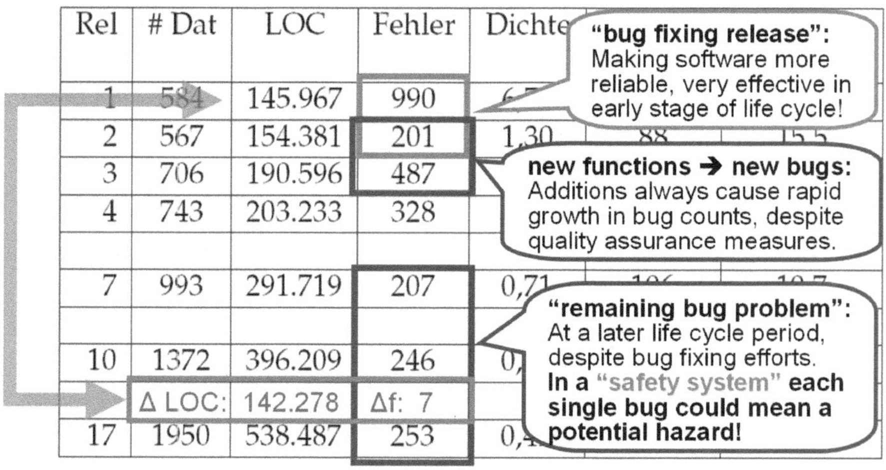

Fig. 4. Bug fixing history and growth statistics of an AT&T communication server software package over a life cycle of 17 releases [9], [13].

Even though that extensive testing has been and still is proposed as an effective method for detecting errors, it has become evident, that by testing alone the correctness of software cannot be proven, because tests can only detect those errors for which the test engineer is looking for [14]. This means ultimately that there is no way to base a safety case on testing alone, because the goal is not to find errors, but to prove the absence of errors. One of the great pioneers of software engineering, Edsgar W. Dijkstra, has put it into the following words [15]:

"Program testing can be a very effective way to show the presence of bugs, but is hopelessly inadequate for showing their absence."

We have even to admit that at the current state of software technology, there is no generally accepted single method to prove the correctness of software that means, there is no way to prove the absence of bugs, at least not for software in a range of 100 TLOC or more. The only promising strategy for minimizing errors is:

1. The development of functional and design specifications (architecture) has to be given top priority and needs adequate resources,
2. The safety part of the software has to be kept as small as possible, and
3. The software life-cycle has to undergo a broad as possible, manifold, and continuous review process.

Successful software projects require for the first point, the specification, at least between 20% and 40% [12] of the total development cost, depending on the size of the final software package. Trying to save at this point will almost certainly result in inflated costs during later project phases (see figure 5).

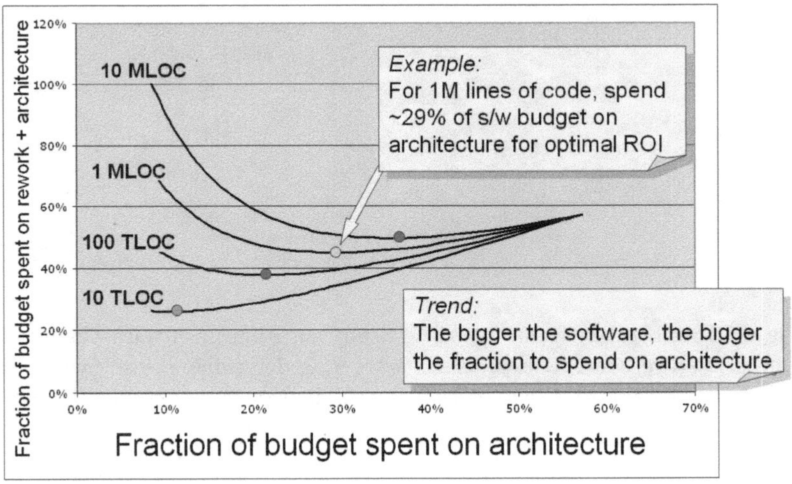

Fig. 5. Fraction of the over all project budgets spent on specification (architecture) versus fraction of budget spend on rework + architecture, which defines a so called "sweet spot" where it reaches its minimum [12]. However this cost function does not take any potential damages into account, which might result from fatalities caused by software bugs.

Formal modeling methods and close communication with the end-user may be helpful in this stage, especially when operational scenarios can be modeled formally as well in order to verify and validate the design. Specification, modeling and reviews by closely involving the customer may even require several cycles in order to come to a satisfactory result.

The second point can only be determined within the project itself by strictly separating safety critical from non-safety related functions and only focusing on pure safety functions in the vital part of the control system. NASA has even suggested a certain maximum limit for code size, related to the quality measures employed, which will most likely cause a "mission failure", according to figure 6. That means, by applying a certain set of quality assurance measures and due to the then reachable bug-density, it does not make sense to start a mission, if the code size has exceeded that particular limit, simply because the failure of the mission would be almost certain. NASA is doing that reasoning usually for a single spacecraft per mission, representing a very high value and involving, at the most, very few human lives.

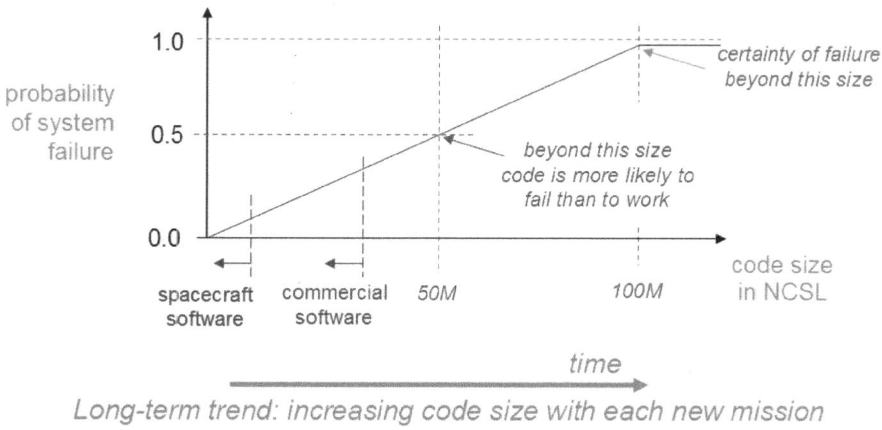

Fig. 6. Graph taken from a NASA "Study on Flight Software Complexity" [12] suggesting a reasonable limit for software size, determine a level, which results in "certainty of failure beyond this size" (NCSL: Non-Commentary Source Lines = LOC as used in this paper).

Large railway operators in the opposite may have several hundreds of trains, representing the same level of material value (but carrying hundreds of passengers) operating at the same time. Assuming that a mission critical failure in software of a particular size may show up once in 1000 years, would mean for a space mission duration of 1 year, a probability for a mission failure of about 0,1% due to software (equal distribution assumed). For the railway operator however, who operates 1000 trains at the same time, having the same size and quality of software on board may cause a "mission failure" event about once a year, making drastically clear that code size matters.

While the third point is very difficult to get implemented within conventional "closed source" software projects, simply because highly qualified review resources are always very limited in commercial projects. Therefore errors are often diagnosed at a late stage in the process. Their removal is expensive and time consuming. That is why big software projects often fail due to schedule overruns and cost levels out of control and often even been abandoned altogether.

Never the less, continuous further development and consistent use of quality assurance measures can result in a remarkable process of "maturation" of software products, which is demonstrated by the fact that in our example in figure 4 the bug density has been reduced by more than an order of magnitude (initially above 6 bugs/TLOC down to below 0.5 bugs/TLOC). On the other hand, in a later stage of the life cycle, due to the continuous growth of the number of code lines, which seem to go faster than the reduction of the bug density, a slight increase of the total number of bugs, can be observed.

Given a certain methodology and set of quality assurance measures on the one hand and a number of change requests to be implemented per release, then this will result in a certain number of bugs that remains in the software. Many of those bugs stay unrecognized forever. However some are also known errors, but their elimination is either not possible for some reason or can only be repaired at an unreasonably high level of cost. The revelation of the unknown bugs can take several thousand unit operation years (number of units times number of years of operation) and must be considered as a random process. That means for the operator, that even after many years of flawless operation, unpleasant surprises have to be expected at any time. In Europe, in a not too distant future up to 50,000 trains will operate with ETCS, carrying millions of passengers daily, plus unnumbered trains with hazardous material. Then the idea is rather scary that in any of those "European Vital Computers" (EVC), the core element of the ETCS vehicle equipment, between 100 and 1000 undetected errors are most likely left over, even after successfully passing safety case assessments and after required authorization has been granted. Even if we would assume, that only one out of 100 bugs might eventually cause a hazard [12], that still means 1 to 10 mission critical defects per unit. Further more, there will be several different manufacturer-specific variants of fault patterns under way.

2.5 New Technologies Have to Have "At Least Same Level of Safety"

According to German law (and equally most other EU states) defined in the EBO (Eisenbahn Bau- und Betriebsordnung: German railway building and operation regulations) any new technology has to maintain at least the same safety level as provided by the preceding technology [16]. Assuming that the more complex ETCS technology requires about ten times more software code than legacy technology like LZB and PZB as an average and given the fact, that PZB and LZB have already reached a very mature stage of the software integrated after almost 3 decades of continues development, then it seams very unlikely, that the "at least same level of safety" can be proven by using the same technical rules and design practices for a relatively immature ETCS technology. In addition, due to less service experience the criticality of deviations from expected reaction patterns are difficult to assess.

This raises the question whether proprietary software combined with a business model that sells the software together with the hardware device, and then - as up to now - will be operated largely without any defined maintenance strategy, might be inadequate for such a gigantic European project. A project eventually replacing all legacy, but well service proven signaling and train protection systems with one single unified, but less service proven technology, especially when independent verification and system validation is only provided at rare occasions by a very limited number of experts ... or ... then again after a "critical incident" has taken place only?

Instead, a broad and continues peer review scheme with full transparency in all stages of the life cycle of ETCS on-board software products would be highly recommended. In particular during the critical specification, verification and validation phases, following the so called "Linus's Law", according to Eric S. Raymond, which states that:

"given enough eyeballs, all bugs are shallow."

More formally:

"Given a large enough beta-tester and co-developer base, almost every problem will be characterized quickly and the fix will be obvious to someone."

The rule was formulated and named by Eric S. Raymond in his essay "The Cathedral and the Bazaar" [17]. Presenting the code to multiple developers with the purpose of reaching consensus about its acceptance is a simple form of the software reviewing process. Researchers and practitioners have repeatedly shown the effectiveness of the reviewing process in finding bugs and security issues [18].

2.6 Changing Business Model for Software: From Sales to Service

Such problems can be solved by changing the business model. Looking at software as a "service" rather than a "commodity" or "product" (in its traditional definition) is justified by the fact that software is growing continuously in size (but not necessarily increasing the value to the user) as shown in figure 4. Over a life-cycle of 17 releases, the total size of that software grew by more than 300%. Furthermore, about 40% of the modules of the first release had to be fixed or rewritten, due to one or more bugs per module. That means only 60% of the original modules were reusable for second or later releases. It is fair to assume that half of the original 60% virtually bug free code had to be adapted or otherwise modified to use it for functional enhancements. This results in not more than 30% of remaining code, which is about 50 TLOC of the original code having a chance to survive unchanged up to release No. 17.

Biggerstaff [19] and Rix [20] suggest that these assumptions might even be too optimistic, as long as no specific measures have been taken in order to support reusability of code. It can be assumed that a potential sales price of all versions would be at the same level, since all versions serve in principle the same functions. That means during its life cycle only 10% (50 TLOC out of about 500 TLOC) of the final code was left unchanged from the first version in this particular example. In other words: 90% of the work (code lines) has been added by software maintenance and continuous further development efforts over the observed life cycle, which can be considered as "servicing the software". In order to make this a viable business, users and software producers have to contract so called "service agreements" for a certain period of time.

This kind of business model can be a win-win for both, users and manufacturers alike. Manufacturers are generating continuous cash flow, allowing them to maintain a team of experts over an extended period of time, dedicated to continuously improving the software. Users in exchange are having guaranteed access to a qualified technical support team ensuring fast response in a case of critical software failures.

Proprietary software makes it mostly impossible for the user to switch software service providers later on, but leave users in a "vendor lock-in" situation with no competition on the software service market. Competition however is the most effective driver for quality improvement and cost efficiency.

Considering commercial, technical, safety and security aspects, the risks associated with complex closed source software should be reason enough for the railway operators to consider alternatives, in particular when a large economic body, like the European Union, defines a new technological standard.

Watts Humphrey, a fellow of the Software Engineering Institute and a Recipient of the 2003 National Medal of Technology (US), has put the general problem of growing software complexity in these words [21]:

"While technology can change quickly, getting your people to change takes a great deal longer. That is why the people-intensive job of developing software has had essentially the same problems for over 40 years. It is also why, unless you do something, the situation won't improve by itself. In fact, current trends suggest that your future products will use more software and be more complex than those of today. This means that more of your people will work on software and that their work will be harder to track and more difficult to manage. Unless you make some changes in the way your software work is done, your current problems will likely get much worse."

3 Proposal: Free / Libre Open Source Software for ETCS

A promising solution for the previously described difficulties could be given by providing an "Open Source ETCS" onboard system, making the embedded software source code and relevant documentation open to the entire railway sector.

"Open Source Software", "Free Software" or "Libre Software" more often called "**Free/Libre Open Source Software**" short: "**FLOSS**" [22], is software that:

1. Can be used for any purpose,
2. Can be studied by analyzing the source code,
3. Can be improved and modified and

4. Can be distributed with or without modifications.

This basic definition of FLOSS is identical to the "Four Freedoms", with which the Free Software Foundation (FSF, USA, [23]) has defined "free software" and is in line with the open source definition formulated by the Open Source Initiative (OSI) [23].

3.1 Public License for an "European" Project

A potential candidate for a license agreement text could be the most widely used General Public License (GPL) or occasionally called "GNU Public License", which has been published by the Free Software Foundation [23]. Because this license text (and several similar license texts as well) is based on the Anglo-American legal system. In Europe applicability and enforceability of certain provisions of the GPL are considered as critical by many legal experts. The European Union has recognized this problem some time ago and has issued the "European Union Public License" text [25], which not only is available in 22 official EU languages, but is adapted to the European legal systems, so that it meets essential requirements for copyright and legal liability issues. The EU Commission recommends and uses this particular License for its own "European eGovernment Services" project (iDABC [26]).

A key feature of the aforementioned license types is the so-called strong "Copy Left" [27]. The Copy-Left requires a user who modifies, extends or improves the software and distributes it for commercial or non-profit purposes, to make also the source code of the modified version available to the community under the same or at least equivalent license conditions, which has applied to the original software. That means everybody will get access to all improvements and further developments of the software in the future. The distribution in principle has to be done free of charge, however add-on services for a fee are permissible. That means for embedded control systems, that software-hardware integration work, vehicle integration, homologation and authorization costs can be charged to the customer as well as service level agreements for a fee are allowed within the EUPL.

By applying such license concept to the core functionality of the ETCS vehicle function as defined and already published in UNISIG subset 026 of the SRS v3.0.0 [1] all equipment manufacturers as well as end-users would be free to use this ETCS software partly or as a whole in their own hardware products or own vehicles. Due to the fact that a software package of substantial value would be then available to all parties, there would be not much incentive any more for newcomers to start their own ETCS software development project from scratch, but would more likely participate in the OSS project and utilize the effect of cost sharing. Also established manufacturers, who already may have a product on their own, might consider sharing in for all further add-on functions by trying to provide an interface to the OSS software modules with their own existing software.

This will result in some kind of an informal or even formally set up consortium of co-developing firms and a so called "open source eco-system" around this core is most likely to evolve. This has been demonstrated by many similar FLOSS projects. The effect of cooperation of otherwise in competition operating firms, based on a common standard core product, is often called "co-competition".

In analogy with other similar "open" projects the name "openETCS" has been suggested for such a project. Occasionally expressed concern that such a model would squander costly acquired intellectual property of the manufacturers to competitors does not really hit the point, because on the one hand the essential functional knowledge, which is basically concentrated in the specification, has already been published by UNISIG and ERA within the SRS and cannot be used as unique selling point. On the other hand implementation know-how for specific hardware architecture and vehicle integration as well as service knowledge will not be affected and has the potential to become part of the core business for the industry. In addition, for the pioneering manufacturer open up his own software could not be better investment money, if this software becomes part of an industrial standard, which is very likely (if others are not quickly following this move) as demonstrated several times in the software industry. Not only that, but since safety related software products are closely related to the design process, tools and quality assurance measures, the pioneering OSS supplier would automatically make his way of designing software to an industrial standard as well (process standardization). Late followers had simply to accept those procedures and may end up with more or less higher switching costs, giving the pioneer a head start. Even in the case that one or two competitors would do the same thing quickly, those companies could form a consortium sharing their R&D cost and utilizing the effect of quality improving feedback from third parties and therefore improving their competitive position compared to those firms sticking with a proprietary product concept. The UNU-MERIT study on FLOSS [22] has shown cost lowering (R&D average of 36%) and quality improving effects of open source compared with closed source product lines.

3.2 ETCS Vehicle On-Board Units with "openETCS"

Software that comes with a FLOSS license and a Copy-Left, represent some kind of a "gift with a commitment", namely as such that the "donor" has almost a claim to receive any improvements made and further distributed by the "recipient". That means all the technical improvements, which have been based on collective experiences of other users/developers and integrated into product improvements need to be distributed so that even the original investor gets the benefits. By recalling the fact that during the life cycle of a large software product, as shown in figure 4, more than 90% of the code and improvements were made after the first product launch, means that sharing the original software investment with a community (eco-system) becomes a smart investment for railway operators and manufacturers alike, by simply reducing their future upgrade and maintenance costs significantly.

Rather than starting to develop a new open source software package from scratch, the easiest and fastest way for a user to reach that goal would be simply by selecting one of the already existing and (as far as possible) service proven products from the market and put it under an appropriate open source license. There are numerous examples from the IT sector, such as the software development tool "Eclipse", the successor to IBM's "Visual Age for Java 4.0.", source code was released in 2001 [28], the Internet browser "Mozilla FireFox" (former: Netscape Navigator), and office communication software "Open Office" (former: StarOffice) and many more.

3.3 Tools and Documents Need to be Included

In the long term it will not be enough only to make the software in the onboard equipment open. Tools for specification, modeling and simulation as well as software development, testing and documentation are also essential for securing quality and lowering life cycle cost. To meet the request for more competition in the after sales software service business and avoiding vendor lock-in effects, requires third parties to be in a position to maintain software, prepare safety case documents, and get the modified software authorized again without depending on proprietary information. A request no one would seriously deny for other safety critical elements e.g. in the mechanical parts section of a railway vehicle, like loadbearing car body parts or wheel discs.
The past has shown that software tools are becoming obsolete quite often due to new releases, changing of operating systems, or tool suppliers simply going out of business, leaving customers alone with little or no support for their proprietary products. Railway vehicles are often in revenue service for more than 40 years and electronic equipment is expected to be serviced for at least 20 years and tools need to be up to the required technical level for the whole period.

The aircraft industry with similar or sometimes even longer product lifecycles has realizing this decades ago, starting with ADA compiler in the 1980th, specifically designed for developing high assurance software for embedded control design projects, originally initiated by the US Air Force and developed by the New York University (GNAT: GNU NYU Ada Translator), which is available in the "public domain" and further developed by AdaCore and the GNU Project [3], [23] and a somewhat more sophisticated tools chain, which is called TOPCASED, initiated by AIRBUS Industries [29].

TOPCASED represents a tools set, based on ECLIPSE (another OSS software development tools platform [28]) for safety critical flight control applications with the objective to cover the whole life cycle of such software products, including formal specification, modeling, software generation, verification and validation, based on FLOSS in order to guarantee long term availability.

TOPCASED seams to be a reasonable candidate for a future openETCS reference tools platform, since it is a highly flexible open framework, adaptable in various ways for meeting a wide range of requirements.

Today manufacturers in the rail segment are using a mix of proprietary and open source tools, since some software development tools like ADA and other Products from the GNU Compiler Collection (GCC) [24] have already been used in several railway projects. Even FLOSS tools, not specifically designed for safety applications, like BugZilla for bug tracing and record keeping, have already been found its way into SIL 4 R&D programs for railway signaling [30].

The importance of qualified and certified tools is rising, since it became obvious, that poor quality tools or even malware infected tools can have a devastating effect on the quality of the final software product. §6.6 of proposed prEN 50128:2009 norm [31], "modification and change control", requires to take care of the software development tools chain and processes, which in the future formally have to comply with requirements for the respective SIL level of the final product.

Recent news about the STUXNET attack, a type of malware (worm) specifically designed to target industrial process control computers via its tools chain (maintenance PCs with closed source operating system) has made pretty clear, that no one can be lulled into security even not with control and monitoring systems designed for safety critical embedded applications [7].

Ken Thompson, one of the pioneers of the B Language, a predecessor of C, and UNIX operating system design has demonstrated in his "Reflections on Trusting Trust" [2] that compilers can be infected with malicious software parts in a way that the resulting executable software (e.g. an operating system) generated by this compiler out of a given "clean" (means: free of malware) source code, can be infected with a backdoor, almost invisible for the programmer. It took several years of research until David A. Wheeler suggested in his dissertation thesis (2009) a method called "Diverse Double-Compiling" [32], based on open source tools for countering the so called "Thomson' Hack". Therefore Wheeler suggests on his personal website:

"'Normal' mathematicians publish their proofs, and then depend on worldwide peer review to find the errors and weaknesses in their proofs. And for good reason; it turns out that many formally published math articles (which went through expert peer review before publication) have had flaws discovered later, and had to be corrected later or withdrawn. Only through lengthy, public worldwide review have these problems surfaced. If those who dedicate their lives to mathematics often make mistakes, it's only reasonable to suspect that software developers who hide their code and proofs from others are far more likely to get it wrong." ... "At least for safety-critical work making FLOSS (or at least world-readable) code and proofs would make sense. Why should we accept safety software that cannot undergo worldwide review? Are mathematical proofs really more important than software that protects people's lives?" [3]

3.4 "Open Proof" the ultimate Objective for openETCS

Wheeler's statement confirms the need for an open source tools chain to cover the software production and documentation process for verification and validation into the open source concept in total, providing an "Open Proof" (OP) methodology [33]. OP should be then the ultimate objective for an openETCS project, in order to make the system as robust as possible for reliability, safety as well as for security reasons. An essential precondition for any high quality product is an unambiguous specification. Until this day only a written – more or less – structured text in natural language is the basis for ETCS product development, leaving more room for divergent interpretation (figure 3) than desirable.

A potential solution for avoiding ambiguities right in the beginning of the product development process could be the conversion into a formal that means "mathematical" description of the functional requirement specification. As recommended by Jan Peleska in his "Habiltationsschrift" (post doctorial thesis) [34]:

"... how the software crisis should be tackled in the future:

- *The complexity of today's applications can only be managed by applying a combination of methods; each of them specialized to support specific development steps in an optimized way during the system development process.*
- *The application of formal methods should be supported by development standards, i.e., explanations or "recipes" showing how to apply the methods in the most efficient way to a specific type of development task. ...*
- *The application of formal methods for the development of dependable systems will only become cost-effective if the degree of reusability is increased by means of reusable (generic) specifications, reusable proofs, code and even reusable development processes."*

Despite the fact that several attempts have been made in the past, a comprehensive Formal Functional Requirement Specification (FFRS) has never been completed for ETCS due to lack of resources and/or funding. Based on proprietary software business concepts there is obviously not a positive business case for suppliers for a FFRS.

Formal specification works does not have to be started from scratch, because there are already a number of partial results from a series of earlier work, although that different approaches, methods and tools have been used [35], [36], [37]. Evaluating those results and trying to apply a method successfully applied in several open source projects and known as a so called "Stone Soup Development Methodology" might be able to bring all those elements and all experts involved together in order to contribute to such project at relatively low cost [3], [38].

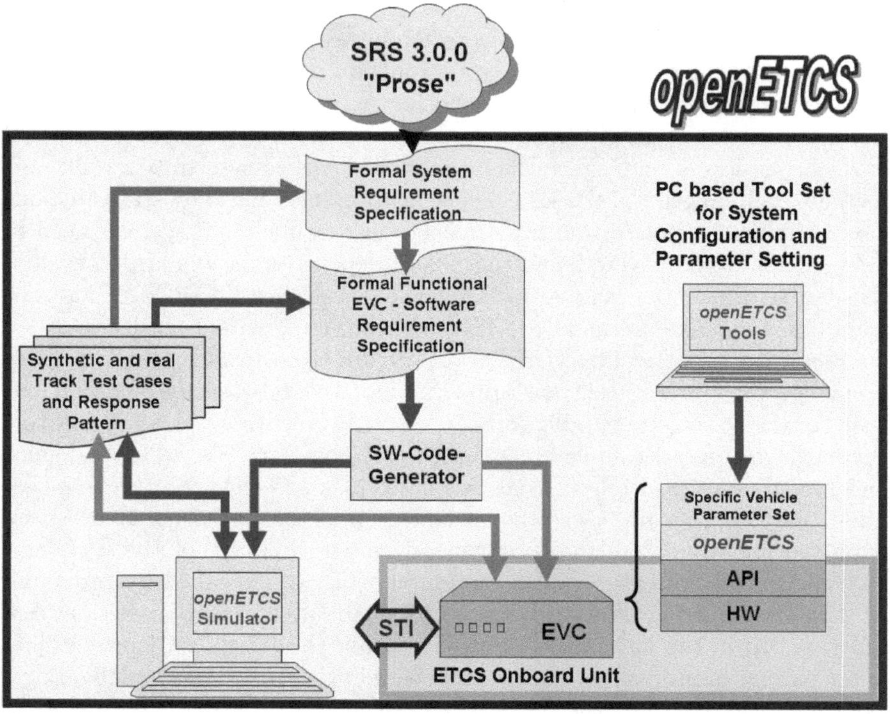

Fig. 7. Proposed openETCS based on ERA's base line 3 SRS natural English specification text (= "prose") converting into a formal functional specification to define software for modeling as well as embedded control integration, providing equipment manufacturers to integrate the openETCS kernel software via API into their particular EVC hardware design.

3.5 Formal Methods to validate Specification for openETCS

In the first step of formalization only a generic, purely functional and therefore not implementation related specification has to be developed. This can be mainly done in the academic sector and by R&D institutes. However railway operators have to feed in their operational experience, in order to make sure that man-machine-interactions and case studies for test definitions are covering real life operational scenarios and not only synthetic test cases of solely academic interest.

For verification purposes a test case data base need to be derived from the functional specification and supplemented by a response pattern data base, which defines the expected outcome of a certain test case. That database needs to be open for all parties and should collect even all real world cases of potentially critical situations and in particular those cases, which have already caused safety relevant incidents. That means this type of formalized database will keep growing and continuously being completed to make sure that all "lessons learned" are on record for future tests.

State of the art formal specification tools do not only provide formatting support for unambiguous graphical and textual representation of a specification document, but provide also a modeling platform to "execute" the model in a more or less dynamical way. This modeling can be used to verify the correctness and integrity of the ETCS specification itself not only statically, but also dynamically. In addition transitions to and from class B systems need to be specified formally as well and that might depend on national rules, even in those cases where the same class B system is used (e.g. for PZB-STMs "hot stand-by" functions are handled differently in Germany and Austria).

Based on a particular reference architecture the resulting formal functional specification can be transformed in a formal software specification and then converted into executable software code. Even without existing real target hardware, those elements can be used to simulate the ETCS behavior and modeling critical operational test cases in a so called "Software-in-the-Loop" modeling set-up. Once the specification of the functionality has been approved and validated, the code generation can be done for the EVC embedded control system. Standardization can be accomplished by providing an Application Programmer Interface (API) similar to the approach successfully applied in the automotive industry within the AUTOSAR project [39] or for industrial process control systems based on open "Programmable Logic Control" (PLC) within the PLCopen project [40] including safety critical systems.

In addition to the software specification, generation, verification and validation tools chain also tools for maintenance (parameter setting, system configuration, software upload services) have to be included in the OSS concept, as shown in figure 7.

3.6 How FLOSS can meet Safety and Security Requirements

For many railway experts, not familiar with open source development methodology, open source is often associated with some kind of chaotic and arbitrary access to the software source code by amateur programmers ("hackers"), completely out of control and therefore not suited for any kind of quality software production.

This may have been an issue of the past and still being in existence with some low level projects, adequate for their purpose. However since OSS license and R&D methodologies concepts have successfully been applied to unnumbered serious business projects, even for the highest safety and security levels for governmental administration, e.g. within the iDABC, European eGovernment Services Project [26] as well as commercial, avionics [29] and military use [24], a concept based on a group of qualified and so called "Trusted Developers" (figure 8) having exclusively access to a so called "Trusted Repository", which on the other hand can be watched and closely monitored by a large community of developers, being able to post bug reports and other findings visible to the whole community, has made this so called "bazaar" process [17] to a much more robust methodology compared with any other proprietary development scheme. According to several research projects, OSS projects in general tend to find malicious code faster than closed source projects, which is indicated for example in the average life time of so called "backdoors", a potential security threat, which might exist in closed source software for several month or even years, while having an average survival time of days or few weeks, at the most, in the case of well managed OSS projects [3], [4], [5], [32]. Figure 8 demonstrates the principle information and source code flow for a typical FLOSS development set-up.

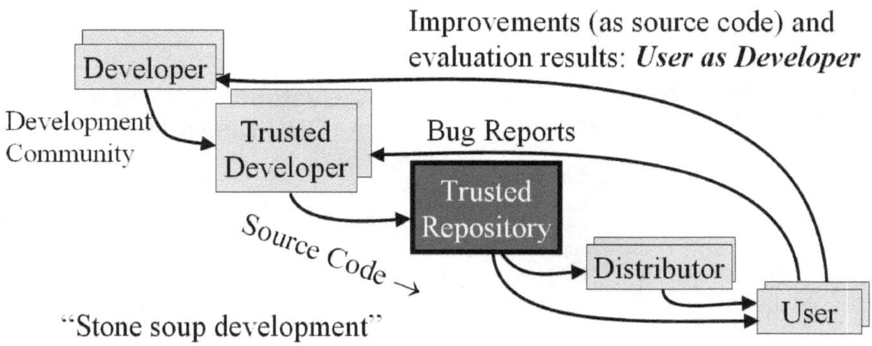

Fig. 8. The classical "Stone Soup Development Methodology" often applied in Open Source Software projects according to [3], where the "User" in most cases is also active as "Developer", which need to be adapted to the rail sector, where "Users" my be more in a reporting rather "developing" role. Only "trusted developers" are privileged to make changes to the source code in the "trusted repository", all others have "read only" access.

It is not in question that well acknowledged and mandatory rules and regulations according to state of the art R&D processes and procedures (e.g. EN 50128) have to be applied to any software part in order to get approval from safety authorities before going into revenue service.

While open source eco-systems in the IT industry are generally driven by users, having the expertise and therefore being in a position to contribute to the software source code themselves, so it seems unlikely for the railway segment to find many end users of embedded control equipment for ETCS (here: railway operators or railway vehicle owners), who will have this level of expertise. Therefore the classical OSS development concept and organization has to be adapted to the railway sector.

Figure 9 shows a proposal for an open source software development ecosystem for openETCS utilizing a neutral organization to coordinate the so-called "co-competition" business model for cooperating several competing equipment integrators and distributors for ETCS onboard products and services based on a common FLOSS standard core module, adapted to the needs of the railway signaling sector providing high assurance products to be authorized by safety authorities (NSA, NoBo).

The concept as shown in figure 9 assumes a license with Copy-Left, requiring in general distributing the source code free of charge, even if code has been added or modified and further distributed, so that the community can re-use the improvements as well. That means that only certain "added values" can be sold for a fee. Typical added values can be service for software maintenance (bug-fixing), software adaptation for specific applications, integration into embedded control hardware and integration into the vehicle system, test and homologation services, training for personnel and so forth.

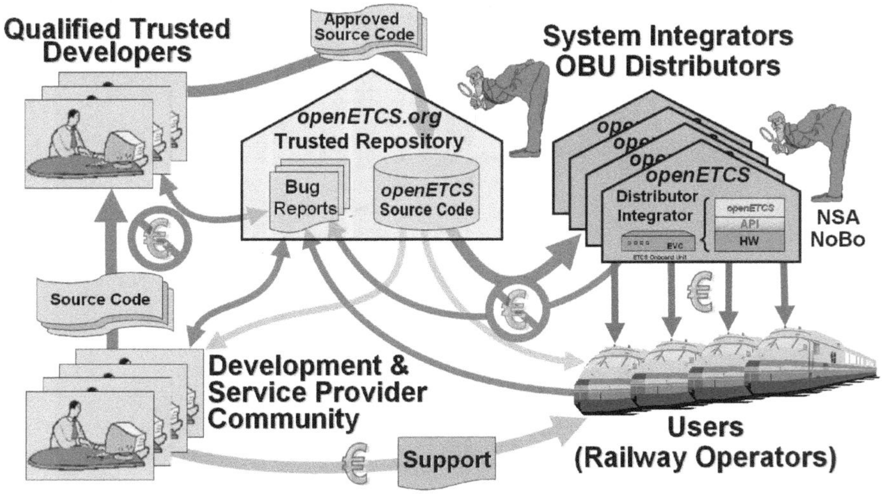

Fig. 9. Proposal for an openETCS eco-system using a neutral organization to coordinate a so-called "co-competition" business model, showing flow of software source code, bug reports and ad-on services provided for a fee.

For further development of the software, especially for the development of new complex add-on functions, costly functional improvements, etc., it might be difficult to find funding, since a Copy-Left in the FLOSS license requires to publish that software free of charge, when distributed.

Therefore many OSS projects are using a so called "dual licensing policy" (or even "multi license policy"), by offering the identical software under two (or more) different license agreements (figure 10).

One might be the European EUPL, a Copy-Left type FLOSS license and the other one can be a "For-Fee-License" (without Copy-Left), which does not require publishing all modification. In exchange ac certain fee has to be paid, which may also provide for warranty and other services. Combined with

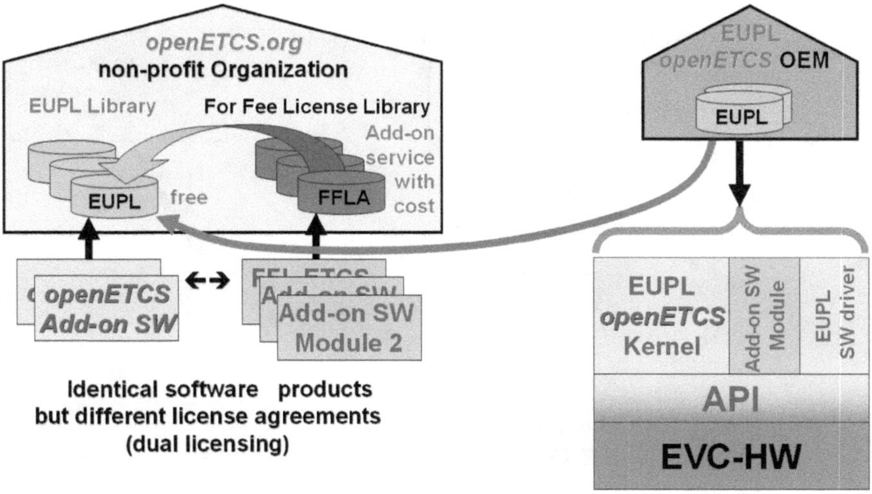

Fig. 10. Dual-Licensing concept providing a cost sharing scheme for financing new functions (Add-on SW Modules) and improvements for those users who need it, by keeping cost low for those not requiring enhanced functionality.

a scheduled release scheme (e.g. defining a fixed release day per year or any other reasonable frequency), all new modules will be available only under the For-Fee-License first, until R&D costs have been paid off by those users, who want to make use of the new functionality, while all others can stick with the older, but free of charge software versions.

Once the new features are paid off, those particular software modules can then be set under the FLOSS license (EUPL). That allows fair cost sharing for all early implementers and does not leave an undesired burden on those users, who can live without the additional functions for a while, but still being able to upgrade later on.

Since those upgrades will be provided by service level agreements through OEMs or software service providers, customers have the choice to either opt for low cost, but later upgrade service or higher priced early implementing services, whatever fits best to their business needs.

The dual-licensing scheme has an additional advantage, allowing even those ETCS suppliers, who are not able, due to technical limitations or legal restrictions caused by their legacy system design or other reasons, to put their software under an OSS license, never the less being able to participate in the cost sharing effects for further add-on functional development. In most cases it is technically much easier to implement a "small API", interfacing just for the add-on functions, rather than providing a fully functional API for the whole kernel (figure 11).

If the non-OSS supplier wants to make use of those Add-on-SW-Modules from the library, he cannot use the OSS-licensed software, but can combine

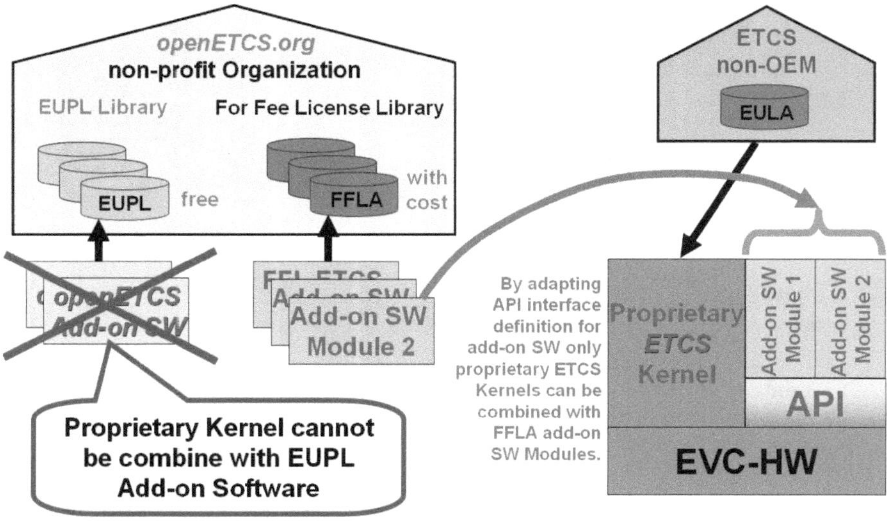

Fig. 11. Applying a dual-licensing model helps even those non-OSS ETCS suppliers participating in cost sharing for add-on modules even if the usage of the openETCS software kernel is not possible due to technical or licensing incompatibilities, but by providing a "Mini-API" some or all future new add-on functions can be integrated.

any proprietary software with alternative licensed software, not including a Copy-Left provision.

Besides commercial matters also technical constrains have to be taken into account when combining software parts, developed for different architectural designs. A concept of "hardware virtualization" has already been discussed to overcome potential security issues [43].

3.7 How to Phase-in an OSS Approach into a Proprietary Environment?

Even though the original concept of the ETCS goes far back into the early 1990 years projecting an open "white-box" design of interchangeable building blocks, independent from certain manufacturers, based on a common specification and mainly driven by the railway operators organized in the UIC (Union International des Chemin de Fer = International Union of Railways), software was not a central issue and open source software concepts were in its infancy [41], [42]. Since then a lot of conceptual effort and detailed product development work has been done, but the "white box" approach has never been adapted by the manufacturing industry.

Despites various difficulties and shortcomings, as mentioned earlier, the European signal manufacturers have developed several products, more or less fit for its purpose and it would be unwise to ignore this status of development and start a brand new development path from scratch. This would just lead to another product competing in an even more fragmented market rather than promoting an effective product standard. In addition, it needs at least one strong manufacturer with undoubted reputation and a sound financial basis in combination with a sufficient customer base to enforce a standard in a certain market. Therefore starting a new product line, by having the need to catch up with more than a decade of R&D efforts is not an option.

Based on this insight, a viable strategy has to act in two ways:

1. Ground work has to be started to provide an open source reference system, based on an unambiguous specification, which means using formal methods, in order to deliver a reference onboard system as soon as possible, which can be used to compare various products on the market in a simulated as well as real world infrastructure test environment. This device needs to be functionally correct, however does not to be a vital (or fail-safe) implementation.
2. At least one or better more manufacturers have to be convinced to share-in into an open source software based business approach by simply converting their existing and approved proprietary ETCS onboard product into an open source software product by just switching to a FLOSS license agreement, preferably by using the European Union Public License (EUPL), including interface definition and safety case documentation. No "technical" changes are required.
3. Once a formally specified reference FLOSS package has been provided, implemented on a non-vital reference hardware architecture, according to step 1, in a future step by step approach all add-on functions and enhancements and future major software releases should be based on formal specifications, allowing a migration of the original manufacturer's software design solution into the formal method based approach, due to the openness of the product(s) from step 2.

Figure 12 demonstrates this two path approach with a conventional roll-out scheme, as planned by a supplier, based on proprietary designs (upper half) and major mile stones for the openETCS project, providing a non-vital OBU based on formal specification and later migrating to a formally specified vendor specific implementation of the kernel software (lower half).

Trying to implement an independent formal open source software package without the backing of at least one strong manufacturer, will most likely fail if no approved and certified product can be used to start with. The only promising way to accomplish the crucial second step in this concept is by using a tender for a sufficiently attractive (large enough) ETCS retrofit project by adding a request for an OSS license for the software to be delivered. The EU commission has provided a guideline for such OSS driven tenders, the so called "OSOR Procurement Guide" (Guideline on public procurement of Open Source Software, issued March 2010, [26]).

As an example, figure 12 shows the time line for an ETCS retrofit project for high speed passenger trains to be equipped by 2012 with an pre-baseline

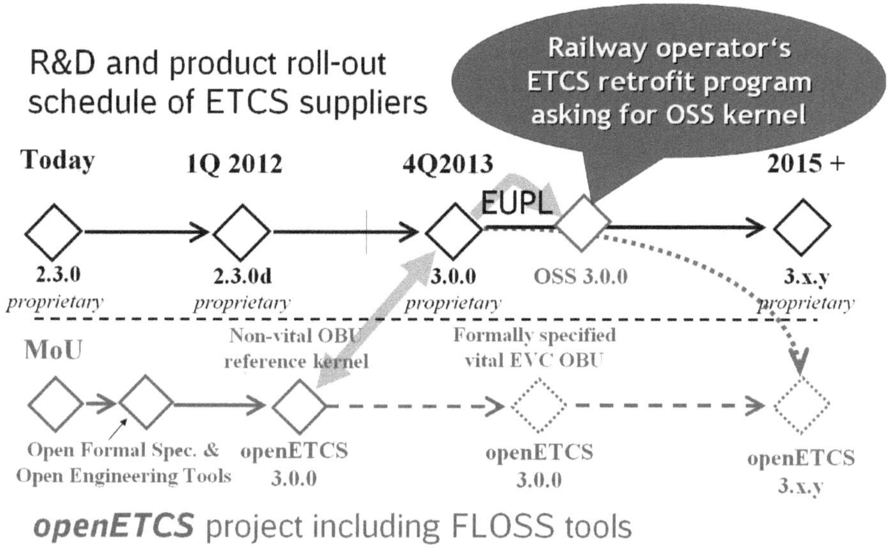

Fig. 12. Interaction between openETCS project providing formally specified non-vital reference OBU for validating proprietary as well into OSS converted industrial products and for future migration to a fully formally specified openETCS software version to implemented in a market product.

3 proprietary software in 2012, to be added by an open source license as soon as the first baseline 3 software package is expected to be released.

3.8 Economical Aspects of openETCS for Europe's Railway Sector

A free of charge, high-quality ETCS vehicle software product on the "market", makes it less attractive, under economical aspects, to start a new software development or even further development of a different but functionally identical proprietary software product. This will lead sooner or later to some kind of cooperation of competing ETCS equipment suppliers, a co-competition with all those suppliers who can and will adapt their own products by providing an API to their particular system.

Due to the fact that very different design and safety philosophies have been evolved in the past years, some of the manufacturers have to decide either to convert their systems or share-in into the co-competition grouping, or otherwise stick with costly proprietary software maintenance on their own.

As figure 4 demonstrates clearly that the increase of the software volume over time may exceed the original volume by a factor of 3. It is unlikely to assume that the development of the ETCS vehicle software will run much differently. Then it will be very obvious that for a relatively limited market, of perhaps up to 50,000 rail cars to be equipped with ETCS in Europe, a larger number of parallel software product development lines will hardly be able to survive.

A study funded by the EU Commission [22] has identified a potential average cost reduction of 36% for the corresponding R&D by the use of FLOSS. As a result, a significantly lower cost of ownership for vehicle operators would accelerate the ETCS migration on the vehicle side.

3.9 Benefits for the ETCS Manufacturers

The core of the ETCS software functionality defined by UNISIG subset 026, to be implemented in each EVC, is a published and binding standard requirement and therefore not suitable for defining an "Unique Selling Proposition" (USP).

As a result it makes perfectly sense from the perspective of manufacturers, to share the development cost and the risk for all R&D of the ETCS core functionality even with their competitors, often practiced in other industrial sectors (e.g. automotive).

Involvement of several manufacturers in the development of openETCS will help to enhance the quality in terms of security and reliability (stability and safety) of the software, because different design traditions and experiences can easily complement each other.

As a FLOSS-based business model can no longer rely on the sales of the software as such, the business focus has to be shifted to services around the software and even other add-on features to the product. That means the business has to evolve into service contracts for product maintenance (further development, performance enhancements and bug fixes). It thereby helps the ETCS equipment manufacturers to generate a dependable long-term cash flow, funding software maintenance teams even long after the hardware product has been discontinued and to cover long term maintenance obligations for the product even by third parties, helping to reserves scarce software development resources for future product R&D.

With respect to the scarcity of well educated software engineers from Universities, FLOSS has the side effect, that openETCS can and most likely will become subject to academic research, generating numerous master and dissertation thesis's and student research projects.

3.10 Benefits for Operators and Vehicle Owners

The use of openETCS is a better protection for the vehicle owner's investment, because an obsolescence problem on the hardware side does not necessarily mean discontinued software service. Modification of the ETCS kernel can also be developed by independent software producers. This enables competition on after-sales services and enhancements, because not only the software sources but also associated software development tools are accessible to all parties.

As shown above, due to the complexity of the software, malfunctions of the system may show up many years, even decades after commissioning. Conventional procurement processes are therefore not suitable, since they provide only a few years of warranty coverage for those kinds of defects. These concepts imply that customers would be able to find all potential defects within this limited time frame, just by applying reasonable care and observation of the product by the user, which does not match experiences with complex software packages with more than 100,000 lines of code.

This finding suggests that complex software will need "care" during the whole life-cycle. Since software matures during long term quality maintenance, means that during early usage, or after major changes, the software may need more intensive care whereas in its later period of use, service intensity may slow down. But as long as the software is in use, a stand-by team is needed to counter unforeseeable malfunctions, triggered by extremely rare operational conditions. As the ETCS onboard software can be considered as "mission critical", operators are well advised to maintain a service level agreement to get the systems up and running again, even after worst case scenarios.

Railway operators and vehicle owners are usually not be able to provide that software support for themselves. They usually rely on services provided by the OEM. However due to slowing service intensity after several years of operation, this service model may not match the OEM's cost structure in particular after the hardware has been phased out. In those cases OEMs are likely to increase prices or even to discontinue this kind of serve. A typical escrow agreement for proprietary software might help, but has its price too, because alternative service providers have first to learn how to deal with the software. Only a well established FLOSS-eco-system can fill in the gap at reasonable cost for the end user, and that is only possible with FLOSS.

DB's experience with FLOSS is very positive in general. For more than a decade, DB is using FLOSS in various ways: In office applications, for the intranet and DB's official internet presence and services on more than 2000 servers world-wide and even in business critical applications. The original decision in favor of FLOSS was mainly driven by expected savings on license cost. However looking back, quality became a more important issue over time, since FLOSS application have had never caused a "service level breach", which cannot be said for proprietary software, selected by applying the same quality criteria. This supports the impression that FLOSS does tend to have a higher quality.

4 Conclusion

The major goal of unified European train control, signaling, and train protection system, ETCS, has led to highly complex functionality for the onboard units, which converts into a level of complexity for the safety critical software not seen on rail vehicles before. A lack of standardization on various levels, different national homologation procedures and a diversity of operational rules to be covered, combined with interfacing to several legacy systems during a lengthy transitional period has to be considered as a major cost driver. Therefore, even compared with some of the more sophisticated legacy ATP and ATC systems in Europe ETCS has turned out to be far more expensive without providing much if any additional performance or safety advantages. Due to ambiguities in the system requirement specification (SRS) various deviations have been revealed in several projects, so that even the ultimate goal of full interoperability has not yet been accomplished. Therefore the development of ETCS has to be considered as "**work in progress**", resulting in many software upgrades to be expected in the near and distant future.

Since almost all products on the market are based on proprietary software, this means a low degree of standardization for the most complex component as well as life-long dependency to the original equipment manufacturers with high cost of ownership for vehicle holders and operators.

Therefore an open source approach has been suggested, not only covering the embedded control software of the ETCS onboard unit itself, but including all tools and documents in order to make the whole product life cycle as transparent as possible optimizing economy, reliability, safety and security alike. This concept is called "open proof" a new approach for the railway signaling sector.

A dual licensing concept is suggested, based on the European Union Public License with a "copy left" provision on the one hand, combined with a non-copy left "for-fee-license" on the other hand to provide a cost sharing effect for participating suppliers and service providers. By offering a trusted repository, a dedicated sources code access policy in combination with a release schedule policy, economical as well as safety and security considerations can be taken into account.

A two step approach, providing a formally specified non-vital reference system and a procurement program, asking for converting existing commercial products from closed source into open source, and later merging those two approaches, is expected to enhance quality and safety parameters in the long run. A neutral independent and mainly not-for-profit organization is suggested to man-age the project involving all major stake holders to define the future product strategy.

The whole openETCS project has to be considered as a business conversion project from a purely competitive sales oriented market into a "co-competitive" service market, enhancing cooperation on standards by enabling competition on implementation and services.

It is well understood that such a change cannot be accomplished even by one of the largest railway operators alone. Therefore several EU railway organizations, as there are: ATOC (UK), DB (D), NS (NL), SNCF (F) and Trenitalia (I) have already signed a Memorandum of Understanding promoting the openETCS concept in the framework of an international project.

References

1. The European Railway Agency (ERA): ERTMS Technical Documentation, System requirements Specification - Baseline 3, SUBSET-026 v300, published 01/01/2010. http://www.era.europa.eu/Document-Register/Pages/SUBSET-026v300.aspx
2. Thompson, Ken: Reflections on Trusting Trust; Reprinted from Communication of the ACM, Vol. 27, No. 8, August 1984, pp. 761-763. http://cm.bell-labs.com/who/ken/trust.html
3. Wheeler, David A.: High Assurance (for Security or Safety) and Free-Libre / Open Source Software (FLOSS); updated 20/11/2009; http://www.dwheeler.com/essays/high-assurance-floss.html
4. Wysopal, Chris; Eng, Chris: Static Detection of Application Backdoors, Veracode Inc., Burlington, MA USA, 2007 http://www.veracode.com/images/stories/static-detection-of-backdoors-1.0.pdf
5. Poulsen, Kevin, "Borland Interbase backdoor exposed", The Register, Jan. 2001, http://www.theregister.co.uk/2001/01/12/borland_interbase_backdoor_exposed
6. EUROPEAN PARLIAMENT: REPORT on the existence of a global system for the interception of private and commercial communications (ECHELON interception system) (2001/2098(INI)), Part 1: Motion for a resolution: A5-0264/2001, 11. July 2001. http://www.europarl.europa.eu/comparl/tempcom/echelon/pdf/rapport_echelon_en.pdf
7. FINANCIAL TIMES Europe: "Stuxnet worm causes worldwide alarm", by Joseph Menn and Mary Watkins, Published: Sept. 24, 2010, Pages 1 and 3 or online version: http://www.ft.com/cms/s/0/cbf707d2-c737-11df-aeb1-00144feab49a.html
8. Schweizerische Eidgenossenschaft, UUS: Schlussbericht der Unfalluntersuchungsstelle Bahnen und Schiffe über die Entgleisung von Güterzug 43647 der BLS AG vom Dienstag, 16. Oktober 2007, in Frutigen. http://www.uus.admin.ch/pdf/07101601_SB.pdf
9. Klaeren, Herbert: "Skriptum Softwaretechnik", Universität Tübingen, Okt. 2007, http://www-pu.informatik.uni-tuebingen.de/users/klaeren/sweinf.pdf
10. McConnell, Steve: Code Complete, 2nd ed. 2004, Microsoft Press; Redmond, Washington 98052-6399, USA, ISBN 0-7356-1967-0
11. Richard H. Cobb, Harlan D. Mills: Engineering Software under Statistical Quality Control. IEEE Software 7(6): 44-54 (1990)
12. Dvorak, Daniel L., (Editor): NASA Study on Flight Software Complexity, Final Report, California Institute of Technology, 2008 Report: http://www.nasa.gov/pdf/418878main_FSWC_Final_Report.pdf Presentation: http://pmchallenge.gsfc.nasa.gov/docs/2009/presentations/Dvorak.Dan.pdf
13. Ostrand, T. J. et al: Where the Bugs Are. In: Rothermel, G. (Hrsg.): Proceedings of the ACM SIGSOFT International Symposium on Software Testing and Analysis, Vol. 29, 2004, Pages 86-96; http://portal.acm.org/ see also: http://www-pu.informatik.uni-tuebingen.de/users/klaeren/sw.pdf (German)
14. Randell, B.: The NATO Software Engineering Conferences, 1968/1969: http://homepages.cs.ncl.ac.uk/brian.randell/NATO/

15. Dijkstra, Edsger W.: The Humble Programmer, ACM Turing Lecture 1972. http://userweb.cs.utexas.edu/users/EWD/ewd03xx/EWD340.PDF
16. Sukale, Margret: Taschenbuch der Eisenbahngesetze, Hestra-Verlag, 13.Auflage 2002
17. Raymond, Eric Steven: The Cathedral and the Bazaar, version 3.0, 11 Sept. 2000 http://www.catb.org/~esr/writings/cathedral-bazaar/cathedral-bazaar/ar01s04.html
18. Pfleeger, Charles P.; Pfleeger, Shari Lawrence: Security in Computing. Fourth edition. ISBN 0-13-239077-9
19. Biggerstaff, Ted J.: A Perspective of Generative Reuse, Technical Report, MSR-TR-97-26,1997,
Microsoft Corporation http://research.microsoft.com/pubs/69632/tr-97-26.pdf
20. Rix, Malcolm: "Case Study of a Successful Firmware Reuse Program," WISR (Workshop on the Institutionalization of Reuse), Palo Alto, CA,. ftp://gandalf.umcs.maine.edu/pub/WISR/wisr5/proceedings/ .
21. Watts S. Humphrey; Winning with Software: An Executive Strategy, 2001 by Addison-Wesley, 1st Edition; ISBN-10: 0-201-77639-1
22. UNU-MERIT, (NL): Economic impact of open source software on innovation and the competitiveness of the Information and Communication Technologies (ICT) sector http://ec.europa.eu/enterprise/sectors/ict/files/2006-11-20-flossimpact_en.pdf
23. Free Software Foundation, Inc.; 51 Franklin Street, Boston, MA 02110-1301, USA: http://www.gnu.org/philosophy/free-sw.html
24. David A. Wheeler: Open Source Software (OSS or FLOSS) and the U.S. Department of Defense, November 4, 2009; http://www.dwheeler.com/essays/dod-oss.ppt
25. European Commission, European Union Public License - EUPL v.1.1, Jan. 9, 2009. http://ec.europa.eu/idabc/en/document/7774
26. European Commission, iDABC, European eGovernment Services; OSOR; Guideline on public procurement of Open Source Software, March 2010, http://www.osor.eu/idabc-studies/OSS-procurement-guideline\%20-final.pdf
27. Wikipedia, terminology: "Copyleft" http://en.wikipedia.org/wiki/Copyleft
28. Eclipse Foundation, "About the Eclipse Foundation",
http://www.eclipse.org/org/\#about
29. TOPCASED: The Open Source Toolkit for Critical Systems;
http://www.topcased.org/
30. Duhoux, Maarten: "Respecting EN 50128 change control requirements using BugZilla variants", Signal+Draht, Heft 07+08/2010, EurailPress http://www.eurailpress.de/sd-archiv/number/07_082010-1.html
31. DIN EN 50128; VDE 0831-128:2009-10; Railway applications - Communication, signal-ling and processing systems - Software for railway control and protection systems; version prEN 50128:2009; Beuth Verlag, Germany, http://www.vde-verlag.de/previewpdf/71831014.pdf (index only)
32. Wheeler, David A.: Countering the Trusting Trust through Diverse Double-Compiling (DDC) 2009 PhD dissertation, George Mason University, Fairfax, Virginia http://www.dwheeler.com/trusting-trust/

33. Open Proof: http://www.openproofs.org/
34. Jan Peleska: Formal Methods and the Development of Dependable Systems, Habilitationsschrift, Bericht Nr. 9612, Universität Bremen, 1996. http://www.informatik.uni-bremen.de/agbs/jp/papers/habil.ps.gz
35. Anne E. Haxthausen, Jan Peleska and Sebastian Kinder: A formal approach for the construction and verification of railway control systems, Journal: Formal Aspects of Computing. Published online: 17 December 2009. DOI: 10.1007/s00165-009-0143-6 Springer, ISSN 0934-5043 (Print) 1433-299X (Online) http://springerlink.metapress.com/content/l3707144674h14m5/fulltext.pdf
36. Lorenz Däubler, Michael Meyer zu Hörste, Gert Bikker, Eckehard Schnieder; Formale Spezifikation von Zugleitsystemen mit STEP, iVA, Techn. Univ. Braunschweig, 2002; http://www.iva.ing.tu-bs.de/institut/projekte/Handout_STEP.pdf
37. Padberg, J. and Jansen, L. and Heckel, R. and Ehrig, H.: Interoperability in Train Control Systems: Specification of Scenarios Using Open Nets; in Proc. IDPT 1998 (Integrated De-sign and Process Technology), Berlin 1998, pages 17 - 28
38. Gary Rathwell: Stone Soup Development Methodology: Last updated December 5, 2000 http://www.pera.net/Stonesoup.html
39. AUTOSAR (AUTomotive Open System ARchitecture); http://www.autosar.org/
40. PLCopen; Molenstraat 34, 4201 CX Gorinchem, NL,: http://www.plcopen.org/
41. UIC/ERRI A200: ETCS, European Train Control System, Overall Project Declaration including the contribution to be made by UIC, Utrecht, NL, Jan. 1992
42. UIC/ERRI A200: Brochure ETCS, European Train Control System, The new standard train control system for the European railways, Aug. 1993, 2nd. Rev. Oct. 1995
43. Johannes Feuser, Jan Peleska: Security in Open Model Software with Hardware Virtualization - The Railway Control System Perspective. Univ. Bremen, 2010 http://opencert.iist.unu.edu/Papers/2010-paper-2-B.pdf

Formal Method and its Application on Train Operation Control System of Chinese high-speed Railway

Tao Tang

Beijing Jiaotong University
ttang@center.njtu.edu.cn

Abstract. With the development of the economic and society of China, a high speed transport method is urgently required to solve the travel problems of people. The train had been speeded up from 120 km/h to 250 km/h in the main line network. Chinese High Speed railway has been rapidly developed. Railway plays a more and more important role in the Chinese transportation systems. For the safety and efficiency of high speed railway, Chinese Train Control system (CTCS) is developed and some new technologies are reseached.
CTCS Level3 is the train control system used in Chinese High speed railway. This system makes use of GSM-R to complete the safe communication between onboard subsystem and RBC subsystem, and employs the track circuit as a backup communicating approach to ensure the safety of the system. Now in China, CTCS Level3 system has already been used in four high speed railway lines, the total length reached 2076 km. And the line from Beijing to Shanghai will be put into operation next year, which also uses CTCS Level3.
To guarantee the safe operation of the train and improve the efficiency of railway traffic, we make use of many formal methods and propose some approaches of modeling, formal verification and developing of the CTCS Level3 train control system, including the Specification Validation and Verification, Hybrid system Modeling and Verification, model based test sequence generation approach and SCADE based safety critical system development method. The advantages of our method are: Establish a track chain among the system specification, model, model checking tools and verification results; generate test sequences automatically and generate safety code automatically.

E. Schnieder, G. Tarnai (eds.), *FORMS/FORMAT 2010*,
DOI 10.1007/978-3-642-14261-1_3,

Towards Open Modular Critical Systems *

András Pataricza

Budapest University of Technology and Economics,
Department of Measurement and Information Systems,
Magyar tudósok krt 2, 1117 Budapest, Hungary
pataric@mit.bme.hu

Abstract. The main strategic objectives of the embedded systems industry are increased design and manufacturing productivity and quality. The appearance of general purpose design and implementation components and platforms requires new paradigms for creating critical embedded systems out of standardized COTS components. Another main driver and simultaneously a source of potential dangers is the integration of embedded systems with the cyber world for value added global services exposing them to security threats. The talk gives an overview on the efforts of supporting the new trends by formal methods complementing model-based design paradigms.

Keywords: Critical embedded Systems, model-based Design, Formal Methods, security-safety co-modeling.

1 Overview

Recent European initiatives, like the ARTEMIS European Technology Platform aim at a drastic increase both in the productivity and quality of embedded systems and the services delivered by them despite the ever growing complexity of the functionality to be implemented. The main means supporting this objective are the widespread use of efficient model-based design methods creating reusable design artifacts and supporting a guaranteed design quality by the design intelligence built into the design frameworks.

This intelligence is able guiding and optimizing the exploitation of the resources provided by standard platforms and pre-manufactured components, while simultaneously guaranteeing the compliance to safety requirements in critical applications. The presentation will give some examples taken from ongoing European projects on the upcoming model-based design for safety methodologies.

Another main development trend in the field of embedded systems is the rapid appearance of cyber-physical systems combining embedded systems

* This work was partly carried out during a visiting professorship at the CASED-Center of Advanced Security Research Darmstadt and supported by the EU projects INDEXYS (ARTEMIS) and MOGENTES (FP7).

E. Schnieder, G. Tarnai (eds.), *FORMS/FORMAT 2010*,
DOI 10.1007/978-3-642-14261-1_4, © Springer-Verlag Berlin Heidelberg 2011

with the Internet-based cyber world in order to create qualitatively new global functionalities.

This synergy appears at first as an upper layer of monitoring, data processing and control. However new opportunities are provided by the evolvability of applications both in the terms of functional and dependability driven adaptation. This way, bringing the formerly pure infrastructure oriented self-* paradigms to embedded systems will become to a reality in the nearest future.

At the same time, the new level of openness absent previously in traditional closed world embedded system designs exposes even critical systems to security intrusions. Formal methods are able co-modeling safety and security faults in order to help robust architectures simultaneously resisting safety and security risks.

1st Day Sessions

2nd December 2010

- Reliability, Availability, Maintainability, Safety and Security (RAMSS) Aspects & Policy
- Modelling of Operations and Maintenance
- RAMS Methods for Railway Operation, Communication and Interlocking
- RAMS Application for Railway Infrastructure and Vehicles

Safety and Security in Transportation Process - Not Just Technical Issue

Margarita Peltekova

University of Transport, 158 Geo Milev Str., Sofia, Bulgaria
mpeltekova@hotmail.com

Abstract. Security these days is global in dimension. It covers issues as climate change, health and the fight against terrorism. The most important components of transportation processes are safety and security of the process and the most important factor for process safety/security is a human factor. Major accidents in transportation process have been attributed to human factor and surprisingly, a human factor is the most poorly investigated aspect of the transportation safety/security process. Improving the human factors design of a process can produce not only improvements in safety, security and health but also gains in quality and productivity in transportation process. Human errors are regarded as one of the main causes for railway accidents these days. In spite of this fact, the consideration of human error probabilities in quantified risk analyses has been very rudimentary. A lack of comprehensive data and analyses lead to the use of estimations and values from other industries. This paper discusses the transferability of human error probabilities for railways and identifies problems in handling methods and values for analysis the security issues concerning any undesirable human behavior and its influence on safety and security of transportation process. The Markov model presented is one proposed solution to the problem. It takes into account both the positive and the negative human impact of violations.

Keywords: Human Factor, Railway Safety, Railway Security

1 Introduction

Railway systems are socio-technical systems. The main requirement for the railway industry is the RAMS(S) EN 50126 standard. The standard demands to take care of all the risks related to RAMS requirements and provides guidance and rules on how to mitigate and prevent such risks. For all countries, the requirement is based on national laws demanding to apply to the RAMS standard. The RAMS standard EN 50126 and EN 50129 are the two standards that form the basis for the safety work, and give important information and requirements for efficient and effective risk management as guaranty for a safe, secure, and reliable transportation process.

E. Schnieder, G. Tarnai (eds.), *FORMS/FORMAT 2010*,
DOI 10.1007/978-3-642-14261-1_5,

A risk analysis process requires studying both technical failures and human errors. Human errors occur when human operators perform their requisite tasks incorrectly or when they perform other non-requisite tasks that could result in degraded system performance. These other tasks may simply enhance the existing uses of a given system, and they can also be intentional deviations from requisite tasks, as is the case with violations. Existing methods for human error analysis generally assess only the unintentional errors that occur during the execution of requisite tasks. In addition, given their heterogeneous results, it would appear that these methods are difficult to use [2]. For this reason, new research is needed to define new and more original methods for explaining and predicting human error, methods that are capable of taking into account the capacity of human operators to both - avoid and correct undesirable events.

One of the proposals dealing with human errors modeling and management introduces the so-called "socio-technical barrier" - a combination of technical, human, and organizational means that prevents or protects against an unwanted consequence. A basic hypothesis would hold that if all physical barriers are operational and all immaterial barriers are respected, the human activity is to be safer.

The transportation systems are integral part of modern society and as such they are vulnerable in every-day life. When in use, the system migrates through the safety boundaries. The new resulting operational space of performance becomes largely positioned outside the initial safe space of performance. This new space is characterized by reduced margins to incidents and accidents (despite that, safety remains acceptable) and numerous violations and deviance. The safety gap should be filled by developing a strategy and implementing new regulations that can fight human violations.

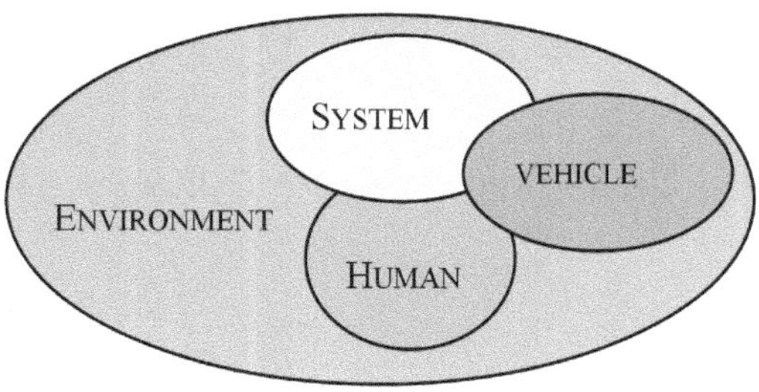

Fig. 1. Vehicle movement factors

The movement of each vehicle is predicted upon the behavioral states of the humans influencing the transportation process, the physical devises encountered by the vehicle during its travel and the local environment. Vehicles interact with these entities and the resulting movement generates a sequence of events that may lead to an unsafe condition. These unsafe conditions result from violations of the safety-critical rules that define safe system performance. The table below illustrates railway accidents in Bulgaria in 10 years period.

Table 1. A91 - Railway accidents - non-official draft from accidents in Safety Data Base - for the period of 1998-2009 year, NRIC, BG

Accidents	Number of accidents	Number of Fatalities
Collisions	52	5
Derailments	15	
Of persons caused by rolling stock in motion	390	201
At level crossing	61	35
Others	2	9

The data presented includes probable-cause information for each accident. The main causes for accidents, especially derailments, were reported to be public behavior and poor maintenance.

2 Modeling Human Behaviour

Human operators are important components of the railway systems, and, as such, can "fail" if their behavior deviates from the behavior stipulated in the system specifications. Clearly, human operators are able to avoid and/or correct incidents or accidents; however, they may also be the cause of such events. Many methods can be used to analyze the potential deviation from the rules, but, regardless of the method chosen, the designer must choose the optimal means of prevention or protection, given the nature of the human error. Thus, to protect a system from potentially unsafe human behavior is as important as to protect the entire transportation process from unsafe railway system behavior.

Human errors are regarded as one of the main causes for railway accidents these days. In spite of this fact, the consideration of human error probabilities in quantified risk analysis has been very rudimentary up to now. A lack of comprehensive data and analysis in literature lead to incomplete estimates and values.

In general, the human errors could be classified in two categories - errors of omission and errors of commission [1,6]. The first implies the lost of one or more steps in a procedure. The last is when a different procedure was made. The main issue in both of them is that the person is unconscious of the error. The models of such type of human errors are made using a well-known techniques predicting human error probabilities and evaluating the degradation of a man-machine system. The basics steps to mitigate and prevent human errors are the following:

- List and analyze human operations
- Estimate the error probabilities
- Estimate the effects of human errors on the system failures
- Estimate working environment

The basic model of human behavior is shown in 2.

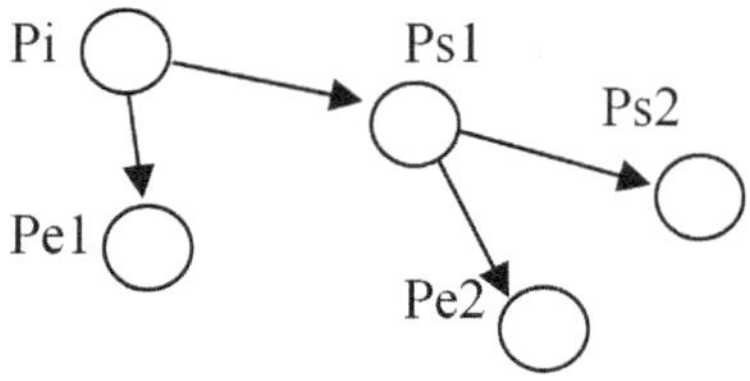

Fig. 2. Human Behavior State Model

Human Error Modeling is a complex task that depends on a set of organizational, qualification and control aspects. Several conditions must be taken into account during the evaluation process: the administrative control, a special qualification for the task, the working environment, and stress. The results depend on component complexity, time and personnel availability, administrative control, supervision and qualification for each personnel, and level of communication between them.

The influence of human behavioral affects transportation process as safety-critical component. Stimuli that have to be recognized include physical and operational environment to which human are exposed. In response to a given stimulus or set of stimuli, humans recognize the need to perform a given action. If such recognition is not made, then an error of omission occurs. As a result, the undesired event succeeds and the accident begins. The sequence can be represented graphically by using the Event Trees built on the basis of statistical information. The branches represent the erroneous actions - Pe and the successful human action - Ps. All probabilities, except those in the first branching, are conditional probabilities.

Other well known techniques to model human error are Petri Nets, Markov Trees, and Fault Trees. Fault Trees do not represent dependencies among human actions. Human behavior can be accurately described as a set of dynamic modes through the dynamic Markov models. Markov chains have been used to model human behavior in controlled or uncontrolled situations. The parameters of the model must be estimated from data. The estimation procedure assumes that the observed process is stationary.

Humans must recognize the need to perform a given action. If such recognition is not made, then an error of omission occurs. Such an error is denoted within Figure 3 by the probability of non-response, S 3. [6] Once the need for an action has been recognized, then the human must determine if the intended action is correct. The ability to detect the presence of inappropriate stimuli and to subsequently correct its error is denoted within this model as "coverage." The probability for coverage is S 4. Once an individual identifies the need to perform the intended action and if this action is performed incorrectly, then an error of commission occurs. The likelihood of such an error is denoted within Figure 3 by the probability of non-compliance, S 6.

The Markov Tree shown has initial state S 1. The human behavior state model shown in Fig. 3 is constructed on the base of statistical data gathered from railway employees and students during practice.

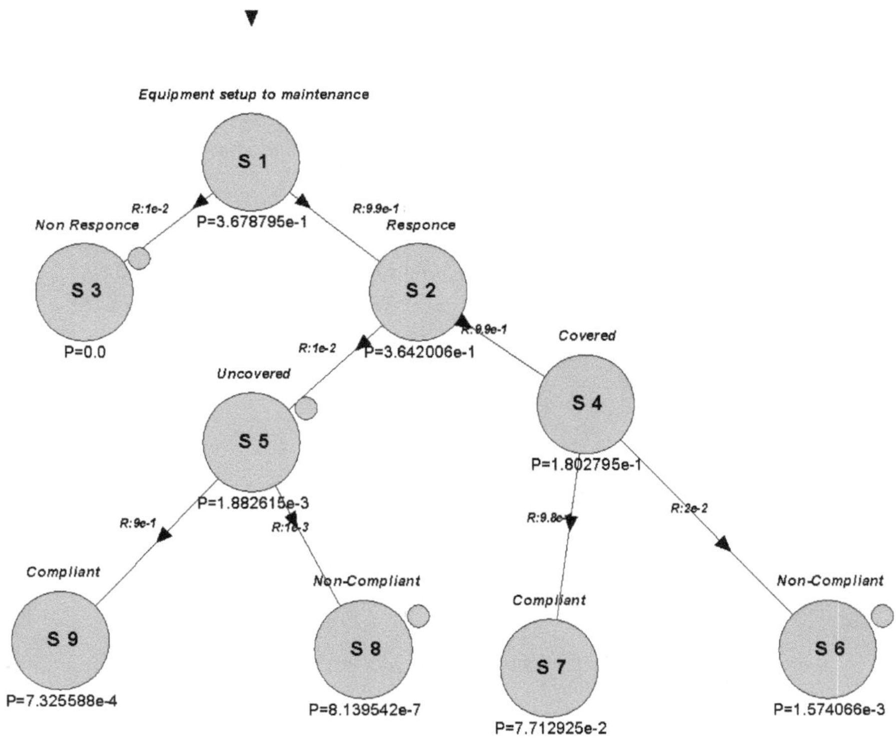

Fig. 3. Human Behavior State Model

The probability of dangerous errors of operators on railway traffic control desk is assumed to be $1 \cdot 10^{-2}$ - $1 \cdot 10^{-3}$ and the probability of correct operation is between 0,8 and 0,95. [5]

3 Modeling the Impact of Human Behavior on Safety and Security of the Transportation Process

Vehicles movement is influenced by the vehicle's interaction with humans and physical devices. The states of the transportation process can be described through the Markov graph as follows:

- P1 - Human behavior and system operation are correct, no violation on critical safety-system states
- P2 - System/human in fail-safe mode
- P3 - Human violation undetected
- P4 - Undetected device failure
- λ - Failure rate
- μ - Repair rate
- C - Coverage
- Pfs - Probability of unsafe human operation
- λDh - Unsafe human violation failure rate

The sequence of events describing the dynamic states of the system is modeled through Markov graph given in Figure 4:

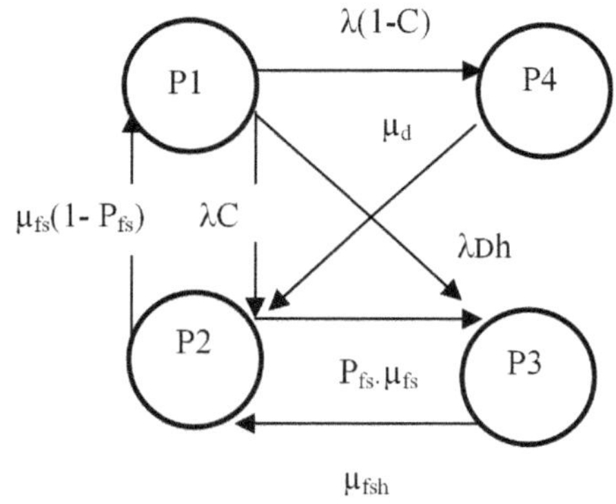

Fig. 4. Railway System Markov Model

The proposed model gives two dangerous system states - probability of unsafe human behavior P3 and well-known probability of dangerous failures of the railway system - P4. For the probability of unsafe human behavior (P3) are considered not only the human errors during maintenance but also intentional workforce, vandalism and unreasonable imprudent unsafe human behavior during working conditions of the system. The probability of the system being in a given state can be found by solving the homogeneous differential equations (1) that describe the Markov process:

$$\frac{dP1(t)}{dt} = -(\lambda + \lambda Dh)P1(t) + (1 - Pfs) \cdot \mu fs \cdot P2(t)$$
$$\frac{dP2(t)}{dt} = -\mu fs \cdot P2(t) + C\lambda fsh \cdot P3(t) + \mu d \cdot P4(t)$$
$$\frac{dP3(t)}{dt} = -\mu fsh \cdot P3(t) + Pfs \cdot \mu fs \cdot P2(t) + \lambda Dh \cdot P1(t)$$
$$\frac{dP4(t)}{dt} = -\mu d \cdot P4(t) + (1 - C) \cdot \lambda P1(t) \quad (1)$$

For steady state probabilities we obtain the following equations:

$$P1 = \frac{\mu fsh \cdot \mu fs \cdot \mu d \cdot (1 - Pfs)}{(\lambda - \lambda Dh) \cdot \mu fsh \cdot \mu d + \mu fsh \cdot \mu fs \cdot (1 - Pfs) \cdot (\mu d + \lambda - \lambda C) + \mu d \cdot \mu fs \cdot (\lambda Dh - \lambda Pfs)}$$
$$P2 = \frac{(\lambda - \lambda Dh) \cdot \mu d \cdot \mu fsh}{(\lambda - \lambda Dh) \cdot \mu fsh \cdot \mu d + \mu fsh \cdot \mu fs \cdot (1 - Pfs) \cdot (\mu d + \lambda - \lambda C) + \mu d \cdot \mu fs \cdot (\lambda Dh - \lambda Pfs)}$$
$$P3 = \frac{\mu fs \cdot \mu d(\lambda Dh - \lambda Pfs)}{(\lambda - \lambda Dh) \cdot \mu fsh \cdot \mu d + \mu fsh \cdot \mu fs \cdot (1 - Pfs) \cdot (\mu d + \lambda - \lambda C) + \mu d \cdot \mu fs \cdot (\lambda Dh - \lambda Pfs)}$$
$$P4 = \frac{\lambda(1 - C) \cdot \mu fsh \cdot \mu fs \cdot (1 - Pfs)}{(\lambda - \lambda Dh) - \mu fsh \cdot \mu d + \mu fs \cdot (1 - Pfs) \cdot (\mu d + \lambda - \lambda C) + \mu d \cdot \mu fs \cdot (\lambda Dh - \lambda \cdot Pfs)} \quad (2)$$

The result from the analysis has shown the importance of human factor for a safe and secure transportation process.

The probability of the system being in a given dangerous state - P3 and P4 in example (Fig.5) shows the positive influence of human-operator on the safety of the system. The model is constructed on the basis of statistical data concerning operator on the railway traffic control desk.

For the probability Pfs (Probability of unsafe human operation), the statistical data shows values from 10^{-1} - 10^{-3} and for λDh (unsafe human violation failure rate) – 10^{-3} - 10^{-6}.

4 Conclusion

Human errors are predictable, and thus can be prevented by changing the design of a system. Psychologists use their knowledge of human perception, response time, and cognition to predict and prevent possible errors. Any organization that professes to have a safety culture should treat human behavior as an important issue. Even the most highly automated systems are designed, installed, and maintained by people. Human error plays a crucial part in most accidents, if not all. In critical systems like transport systems, safety measures against human errors play a substantial role. Human error can be committed in different phases of the life cycle, namely, during system specification and development; and in the longest phase of the life-cycle during operation. In railway operation, several safety-critical tasks are assigned to the operators and are not controlled by signaling and interlocking systems. Many tasks are necessary in situations occurring very rarely. Since these situations are unfamiliar to the operators, the demand for good interaction is high and

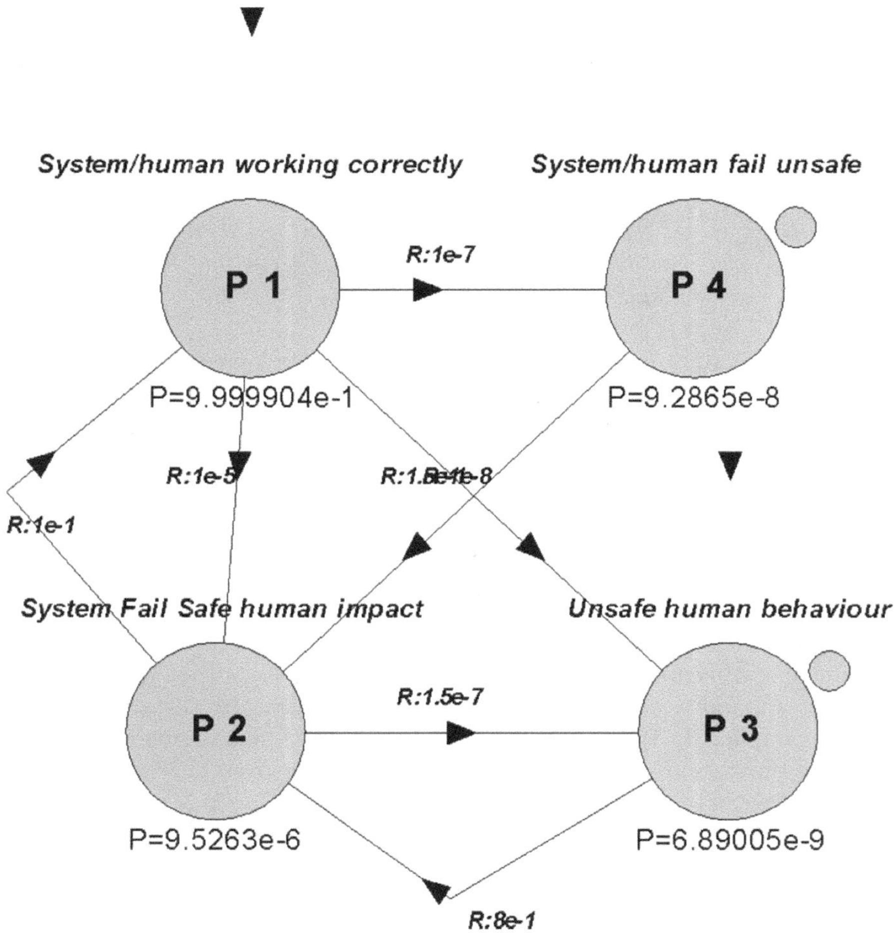

Fig. 5. Example application

pre-training is necessary to provide information and practice to solve these situations. A human agent may communicate with another human agent, and a machine may pass information to another machine in a distributed system. Train control systems have associated accident risks from non-human failures (i.e., mechanical, electrical, and electronic, materials) as well as they have to associate from human failures. Therefore, there is a need to develop an approach for assessing the human failures in train control systems of different type, and to be able to estimate the probabilities of these failures.

References

1. L. Schnieder; E. Schnieder; T. Ständer, Technische Universität Braunschweig, Institute for Traffic Safety and Automation Engineering Langer

Kamp 8, 38106 Braunschweig, Germany, Railway Safety and Security - Two Sides of the Same Coin?!
2. Amit Kumar, P. K. Sinha Mechanical Engineering Department, National Institute of Technology, Patna-800005(BIHAR),India, Human Error Control in Railways
3. Andrew Rae, System Safety and Quality Engineering, 11 Doris St Hill End Queensland 4101, Helping the Operator in the Loop: Practical Human Machine Interface Principles for Safe Computer Controlled Systems
4. Jorge E. Núñez Mc Leod and Selva S. Rivera Human Error Management Optimization in CAREM NPP, Proceedings of the World Congress on Engineering 2009 Vol I WCE 2009, July 1 - 3, 2009, London, U.K.
5. Хр. Христов, Основи на осигурителната техника, София 1990, България russian
6. L. M. Kaufman, Ted C. Giras, Simulation of rare events in transportation systems, Center of Safety-Critical Systems, University of Virginia,
7. Хр. Христов, Електронизация на осигурителната техника, София 1984, България
8. Хр. Христов, Електрически централизации, София 1978, България
9. Е. С. Венцель, Л. А. Овчаров, Теория вероятностей, Москва 1969 russian

The Policy of applying RAMS to evaluate Railway Signalling Systems for reliable Transportation

Kazue Yasuoka[1], Atsushi Watabe[1], Tetsunori Hattori[2], and Masayuki Matsumoto[1]

[1] Electrical & Signal Network Department, East Japan Railway Company, 2-2-2, Yoyogi, Shibuya-ku, Tokyo, Japan
{k-yasuoka,a-watabe,m-matsumoto}@jreast.co.jp
[2] Railway International Standards Center, Railway Technical Research Institute, 2-2-2 Yoyogi, Shibuya-ku, Tokyo, Japan
hattori@rtri.or.jp

Abstract. This paper describes how we adopted RAMS [1] [2] for railway signalling system management. We wanted to use this standard to railway signalling system to improve performance, maintenance and reduce failures. We evaluate railway signaling system by a parameter, and the parameter that shows the customer impact level from a railway signalling equipment system down is selected for evaluating reliability and the risks are analyzed. Then certain problems became clear and countermeasures examined, some of which were improved.

Keywords: RAMS, Signalling System, Reliability Management

1 Introduction

Some railway signalling equipments, such as signals and switch machine, is installed on the site. Other systems, such as interlocking systems, are located indoors. This equipment should be maintained regularly but if failure occurs with these devices, it is necessary to try to find the cause and repair it as quickly as possible. On the other, essential parts from equipment design to end of service life are written in the RAMS standard enacted in 2002.

In this paper, we examined the way of RAMS in railway signalling systems because we wanted to use this standard to achieve better performance, improvement of maintenance and reduction of failure.

2 About East Japan Railway Company

East Japan Railway Company [3] became a share-holding company and is one of the seven railway companies formed on April 1st, 1987, following the

E. Schnieder, G. Tarnai (eds.), *FORMS/FORMAT 2010*,
DOI 10.1007/978-3-642-14261-1_6,

division and privatization of the Japanese National Railways. In October 1993, the company placed its shares on the stock market.

There are 1,705 stations on our 7,527 kilometers network in eastern Honshu, including the Tokyo metropolitan area, and the company serves more than 16 million passengers daily.

Among the many lines operated by JR East, the Shinkansen and Tokyo Metropolitan Area lines have require higher-density, safety and reliability.

Fig. 1. Area served by East Japan Railway Company

3 Evaluation of current railway signalling system by RAMS

3.1 What is RAMS?

RAMS (Railway applications - specification and demonstration of reliability, availability, maintainability and safety) standard is the technology for evaluating safety and reliability of the overall system, and was enacted as an IEC international standard (IEC 62278) in 2002.

What should be carried out in 14 stages from the conceptual to the decommissioning and disposal are defined in RAMS, as shown in Fig.2.

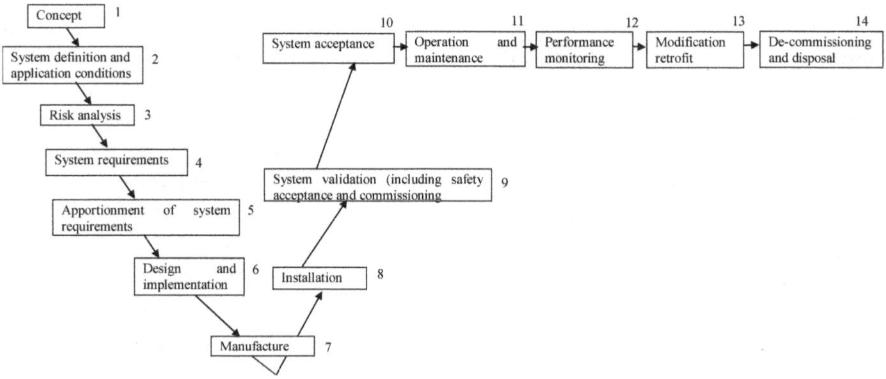

Fig. 2. The "V" representation

By implementing the requirements of RAMS at each stage of development and managing risks in consideration of both the frequency of hazardous occurrences and their severity, it becomes possible for the system to be maintained effectively and economically.

When new system is to be set up on site (tenth cycle of RAMS), test data of a variety of systems, sometimes a report of safety property verification by a third person and risk analysis at a time of design by supplier are verified.

The risk analysis data which we receive before setting up on site is the 'designed value'. The risk evaluation mechanism when we operating is necessary as a system user and we think that risk analysis of RAMS can be applied for operating systems as a method of evaluating reliability [4] [5].

3.2 Problems in failure verification related to railway signalling system in JR East

Meetings that deal with break down are held once a month with each supplier, in which the cause of failure is specified and measures taken with all related equipment when equipment breaks down. That is to say, in our present system of maintenance, we tend to repeat the RAMS stages 11 (Operation and maintenance) to 13 (Modification and retrofit), according to need. When a problem occurs, we often merely go back to the 13th stage and make a repair (see Fig.1, Cycle A).

But when looking at the overall railway signalling system, we need to evaluate the appropriate reliability level of equipment over the entire system for achieving an integrated level of reliability. It is not enough to verify each equipment failure. By starting from the evaluation of the entire system, it can become clear which equipment has problems.

3.3 Importance of reliability management

It is necessity to analyze risk in the course of improving safety and putting those analyses into writing at each stage of the life cycle.

Risk management is concerned with safety. (Risk=Severity level of hazard consequence, not frequency of train delay occurrence.)

In the analysis of risk through RAMS, risk management is carried out by finding safety risks from "frequency of occurrence" and "resulting severity".

We apply a reliability level instead of a risk level to the ongoing systems' evaluation.

We believed that this applied to the stability of train service.

When applying RAMS analysis to equipment reliability, one may consider "total delay time" (the sum total of delays in operation resulting from defective equipment) as a parameter indicating "resulting severity".

We have also considered passengers complaint as an appropriate parameter.

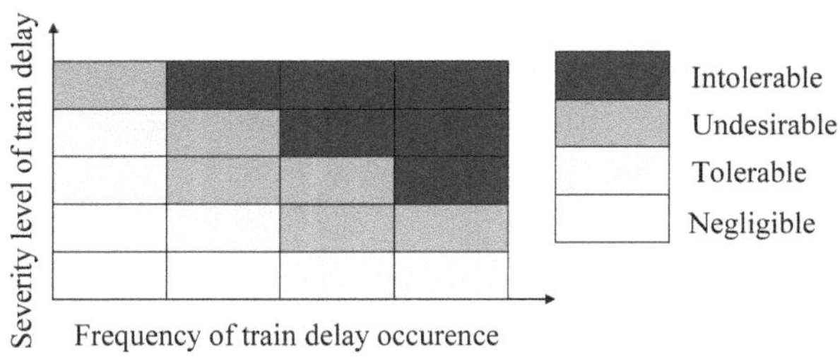

Fig. 3. Example of reliability evaluation and acceptance.

3.4 Method of analysis

We analyzed the data of malfunctions of signalling operations over the last three years within the JR East service area.

Systems down such as those which occur when a system stops working do not have as much impact thanks to the dual system. If one of the two systems down, the other system continues working and thus does not inconvenience customers.

1. First step.
   The failure data for the past three years was classified as shown by the following table.

2. Second step.
   "Total delay time" and "failure rate" are selected as appropriate parameters that show level of severity.
   The reason "total delay time" is selected is that "gtotal delay time" shows the level of impact on customers indirectly. The number of trains is greater when there are many customers, and as a result, total delay time to trains will increase when failure occurs.
3. Third step.
   The following values are calculated from failure data.
   Failure rate = 1/MTBF (1/h)
   MTBF = "total operation time"/"total number of system down number"
   "Total delay time" is divided by number of devices to facilitate comparison according to area.
4. Fourth Step.
   Our target for a 50 km radius from Tokyo Station is to keep delay time under sixty minutes. "Total delay time" that satisfies this condition is set as the "target total delay time". The target total delay time is around 1,100 minutes.

Table 1. Classification of failure data

Item	Classification
Area	50 km radius, 100 km radius from Tokyo Station and other areas.
Equipment type	CTC/PRC [3] , electronic interlocking system, cable, track circuit and etc.
System down factor	man-caused failure, design and manufacture and constant failure

Table 2. Quantity of subsets of signalling systems

Subset of signalling system	CTC/PRC	Track circuit	Cable	Electronic interlocking system
Quantity within 50km radius from Tokyo	153 sets	2,811 sets	5,750km	123 sets
Quantity within 100km radius from Tokyo	162 sets	2,971 dets	6,088km	130 sets

3.5 Analysis result

1. Analysis of each area
   We analyzed the data of signalling operation malfunctions in each area. Consequently, it becomes clear that the impact on customers is greater in an area within a 50 km radius of Tokyo (see fig.4). Up to now, we have concentrated on improving more trouble-free equipment in the 100 km radius around Tokyo Station, but have realized that it is more cost-effective to bring the radius down to 50 km, and that a greater number of passengers enjoy the benefits.

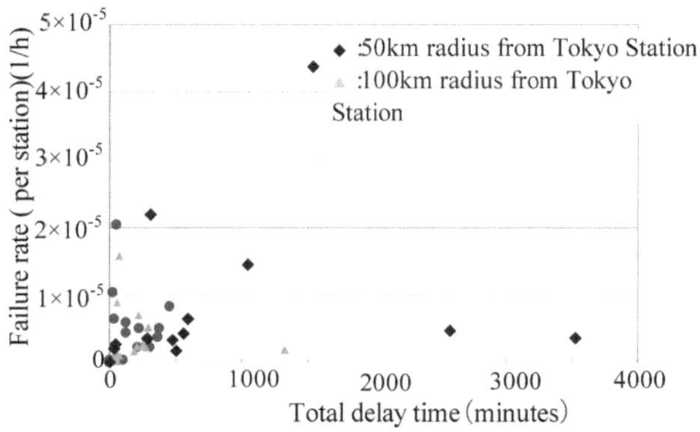

Fig. 4. Analysis in each area

2. Analysis of cause of system downs within 50 km radius from Tokyo.
   It is clear that there is more serious impact within a 50 km radius from Tokyo Station than in other areas. Failure related to cables and CTC/PRC has greater effect on customers but the probability of occurrence is low. In addition, electronic interlocking systems have a higher probability of breaking down, as well as having a greater impact on customers.
   (a) CTC/PRC system down.
   Failure causes of CTC/PRC are analyzed and the frequency of software bags is greater.
   (b) Cable breakdown.
   For cables, "man-caused failure" such as wiring mistakes related to construction is the main cause of breakdowns. "Design and manufacture" is mistakes in connecting.
   (c) Electronic interlocking system.
   For electronic interlocking systems, "constant failure" has a higher occurrence, such as transitory communication abnormalities.

Main measures policy. Based on these analyses, we examined which stage of RAMS is appropriate to return to when measures are taken.

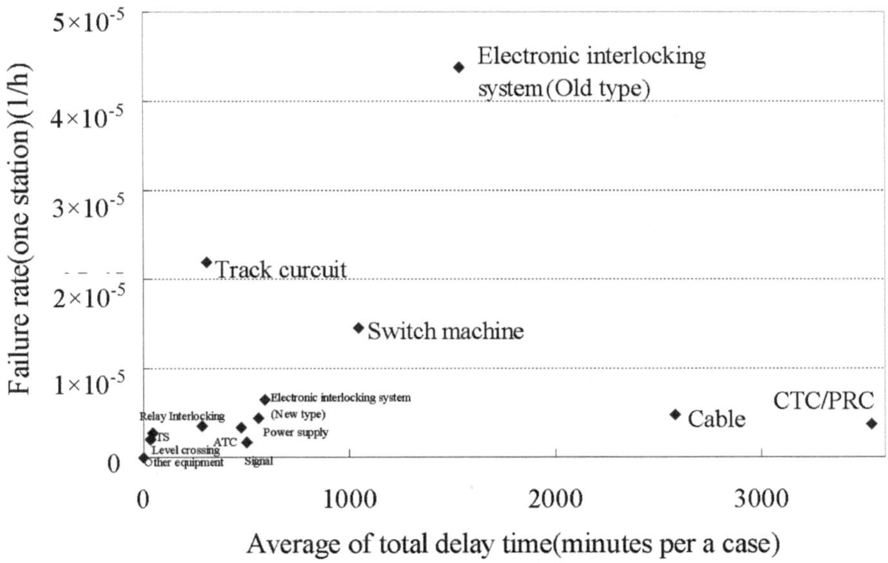

Fig. 5. Cause of system down within 50 km radius from Tokyo Station

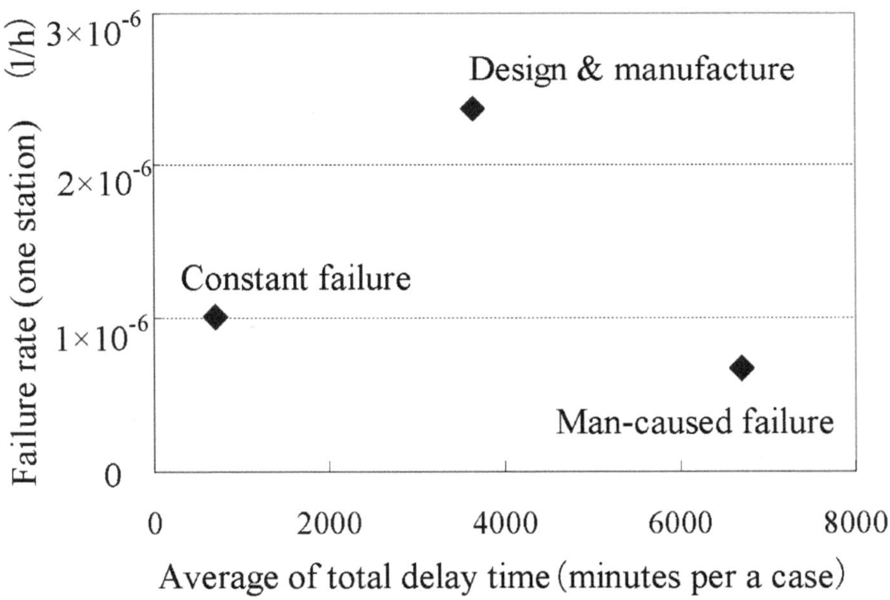

Fig. 6. Analysis of cause of CTC/PRC system down

1. Man-caused failures (CTC/PRC and cable).
   An immediate measure is to train workers to be able to speed up restoration work. However, we would improve human interface and reduce the

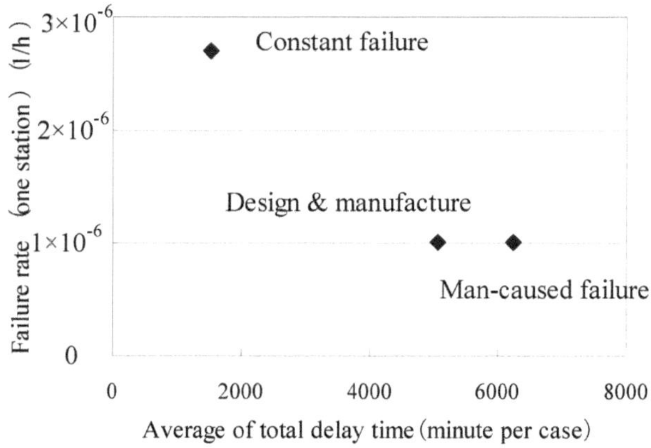

Fig. 7. Analysis of cause of cable breakdown

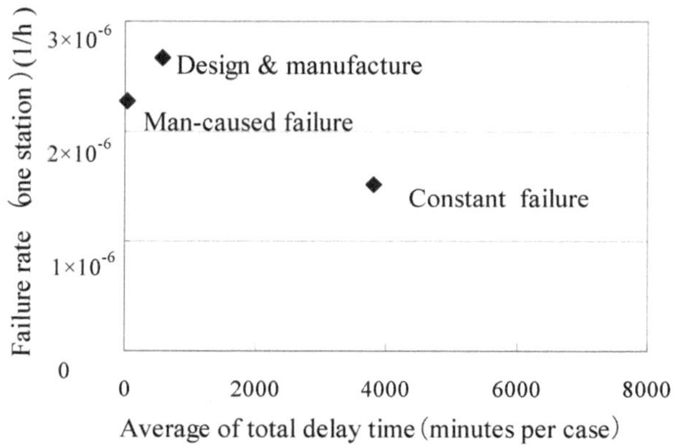

Fig. 8. Analysis of electronic interlocking system breakdown

number of functions when the time comes to replace the equipment. Moreover, equipment would be changed for less likelihood of mistakes, such as reduction of wiring work.

2. Design and manufacture (CTC/PRC and cable).
   In addition to taking measures, in factory pre-shipment tests, reviews of control technique was conducted to prevent system down caused by unpredictable software bugs (3rd cycle of RAMS).
   In addition, since making connections on site may be a factor in manufacturing mistakes, processing on the site has been stopped.
3. Constant failure (electronic interlocking system and cable).
   It is difficult to prevent a transitory error by a component before failures occur. For an electronic interlocking system, the duplex system is standard. We examined control techniques when the system down occurred (3rd cycle of RAMS).

For cables, when we construct them, we include accident preventive measures, such as covering cables with rubber pads where they go under rails.(8th cycle of RAMS).A new signalling system [6] which is a complete duplex system, including the optical cables, has already been developed and will be placed in a 50km radius from Tokyo Station in the near future (1st cycle of RAMS).

4 Conclusion

In this paper, the railway signalling system was evaluated by using RAMS for reliability evaluation of railway transportation.It became possible to set appropriate priorities of measures by evaluating the overall railway signalling system as there is equipment with failures that have only a small impact even if probability of occurrence is high. Moreover, we could analyze not only system down of equipment unit but also man-caused failures.

As a result, it became possible to take more effective measures. Signal equipment will be continuously evaluated by RAMS technique in the future and we will make efforts to set up appropriate conditions. In that case, we should return to various stages of RAMS and measures should be taken.

References

1. Railway applications - Specification and demonstration of reliability, availability, maintainability and safety, IEC 62278 Ed. 1.0:2002 / EN 50126(1999)
2. S.Hiraguri: International Standard overseas railways. Trend and problem concerning RAMS standard In: Railway and electrical technology, Vol. 16, Reliability Engineering Association of Japan, Japan (2005)
3. East Japan Railway Company Information, `http://www.jreast.co.jp/e/index.html`
4. T.Nanya: Fault tolerant computer (in Japanese),Ohmsha,Japan (1991)
5. H.Makabe : Introduction to reliability engineering(in Japanese), Japanese standard association (1990)
6. Y.Hirano,et.al: Development of Railway signalling System based on Network Technology, Proc.of IEEE SMC(2005)

Complementarity between Axle Counters and Tracks Circuits

Marc Antoni

SNCF, Direction of Infrastructure,
Head of the Technologic Innovation Department, Paris, France
Ing. CNAM & SUPELEC, Dr.Ing TUBS, FIRSE, SNCF Expert
(+33)1 53 42 22 52,
marc.antoni@sncf.fr

Abstract. The infrastructure manager choice of type of the track vacancy detection system requires to be based on a safety and signalling functions analysis and not on a dogmatic position. The paper aims at showing that the two main families of these systems are complementary, each of them responding to particular needs in regard to the route category, the traffic duty and the accepted risk.

Keywords: Axle counter, Track circuit, System Safety Analysis, Level for Digital Systems, Railways Critical Systems

1 Introduction

The development of any signalling system uses a whole of "track inoccupancy detectors". These functions have to be developed at the SIL4 level. Two main technologies are used today: axle counters (AC) and track circuits (TC). The other detection modes generally suffer from lower accuracy, reliability and speed. Any experts have suggested retaining one of these two solutions as the unique target system in the Euro-pean standards (TSIs). The aim of the paper is to show why the decision to impose a unique solution in the 'target system' must not be taken without an in-depth systems approach. This analysis has to consider the signalling functionalities, the maintenance principles and the risk acceptance of each country. It seems useless to oppose the two systems: they are complementary and not interchangeable. It is a fact that the signalling functions (including degraded mode, safety rules...) are closely dependant on the nature of the sensors, the change of which can lead to unsafe situations. The infrastructure managers use comprehensive inspection procedures to prevent disruptions in operation and hazards due to defects in the track superstructure. The corresponding procedures are specified in the rules and regulations. The considered necessary effort for the inspections is selectively adapted in accordance with the track duty in terms of traffic density, axle load and speed, and the type of track vacancy detectors. It is naturally greatest on high-speed lines and predominantly characterized

E. Schnieder, G. Tarnai (eds.), *FORMS/FORMAT 2010*,
DOI 10.1007/978-3-642-14261-1_7, © Springer-Verlag Berlin Heidelberg 2011

by shorter inspection intervals. Rail defects arise from a wide range of root causes, up to and including rail breakage. On CWR modern tracks, most of broken rail (BR) situations can be safely detected via track circuits (in France, a BR presenting 2 Ω serial resistance is detected by actual HSL track circuits). Moreover, in many countries, such BR has to be normally reported by the locomotive driver to the operating department via the on-board train radio link today. Generally, the entire set of rail maintenance guidelines does make a distinction between lines with axle counters and lines with track circuits detecting BR: the inspection effort required is consequently greater in the first case. It's an infrastructure manager's choice with a probabilistic or deterministic safety points of view and a global life cycle cost and in regard with the traffic of the lines.

2 Technical presentation of the TC and the AC

Nowadays, the systems of train detection are very important products, absolutely essential to ensure the safety of the railway exploitation. The two basic principles of the detection and localization of trains are track circuits (TC) and axles-counter (AC).

Track circuit principle: a transmitter emits in the track through the two rails an electric coded current which is received at the other end of the track section by a receiver. The track circuit contributes effectively to the train detection and the safety of their movements. In many countries, such as Belgium, France, Great Britain, OSJD countries, Korea the United States or Japan, track circuits are the predominant sys-tems used for train detection providing the signalling system with the locations on the network of any train, and any railway vehicle, it any moment and in full safety. The TC length domain, determining the accuracy of the detection, ranges from 22,5m to 3km, for example France. In most of the countries listed above, HSL track circuits don't need insulation joints in the direct way and their proper functioning is checked by commercial trains using a ground to board transmission. Broken rail detection is as well a safety function ensured by the TC[1], in particular on high density lines. Some TC[2] makes it possible to detect safely and localize a broken rail in a 30km section.

Axle-counter principle: axle-counters compare the number of axles entering and leaving a considered track section. From this calculation results that at any moment the device associated with the track section knows the number of axles present in the zone, to avoid having more than one train at the same time between the two detec-tors. If the in/out difference number takes any value other than zero (1 in some countries), the track section is considered as "occupied"; if the number of axles added up for the track section is null then, obviously, the track section is considered as "free". The detection points count

[1] In any countries it's an safety obligation to detect BR (France, Belgium ...), it's not the case for any other counties (Germany, UK...)

[2] Designed and protected by the SNCF Technological Innovation Department

the number and direction of axles passing the rail contacts and transmit this data to the ACE who defines the section to be "free" or "occupied". In case of an equipment failure, the section will be treated as "failed".

The reset of the systems: The track circuits have no need of resetting function. This property is taken into account in the line traffic organization, in the exploitation and maintenance prescriptions of the countries using the TC and in the interlocking & block design. The reset is required to clear an axle counter section initially for commissioning or when it has become disturbed, i.e. when there is no train in the section but the axle counter indicates occupied as a result of a disturbance. Resetting axle counter sections is a safety-relevant operational procedure, which must be clearly defined in the operator's and maintainer's regulations.

3 Functional differences and specific Constraints

Detecting the inoccupancy of a track section is a safety-critical function. It must therefore be performed with a high level of reliability, consistent with the reliability levels of the other safety functions. Choosing a particular technology for this function necessarily impacts on the whole system, and especially the following:

- the methods and means of operation and maintenance of the fixed plant;
- the design of the infrastructure and signalling installations;
- how safety is achieved in nominal operation and in degraded mode;
- the interoperability-related aspects in the sense of the rolling stock-infrastructure interface, by the imposition of constraints on the rolling stock.

From the standpoint of operation:

It is necessary to consider specific constraints brought by the TC, in particular that:

- the maintenance vehicles and lorries are automatically detected. It has to be considered for the exploitation;
- the transit of some particular vehicles on some less circulated and not electrified track sections ($< 10000T/day$) can lead to the application of specific proce-dures;
- the functioning of the automatic blocks systems and level crossings are not monitored by the operators ...

It is necessary to consider specific constraints brought by the AC, in particular that:

- it imposes the absolute block (one train between to detectors), as it exhibits un-safe failure modes in "permissive blocks", which is used in France and Belgium (UK and France use recently AC of secondary lines with absolute block);

- it does not allow sectional release in the signal boxes (systematically used in France for example). Each switch has consequently its own track circuit and can been released and moved immediately after the train movement[3];
- it requires the train controller to apply special procedures in case of malfunc-tion, including on open track (resetting and verification that the track sections are clear of all traffic). This means that the information train controllers need to follow those procedures;
- it does not allow fulfilling some of the national signalling functions on the French network, such as continuous train warning at level crossings ...

Accordingly, the TC's replacement by ACs, keeping all else constant, would inevitably lead to hazardous situations.

From the standpoint of infrastructure management and maintenance, It is necessary to consider specific constraints brought by the TC, in particular that:

- circuit of the traction return current has to be ensured compatibly with the TC functioning and the track insulation;
- track insulation has to be guaranteed (typically minimal guarantied value from 2,5 to 8 Ω.km in France);
- insulation[4] joints has to be maintained in relation with the traffic load;
- rolling stock has to fulfil some "EMC requirements", in particular regarding the traction return current ...

It is necessary to consider specific constraints brought by the AC, in particular that:

- it seems less simple to maintain, whereas it makes it necessary for maintenance workers to follow special procedures, drawing on railway resources, since AC failures often require manual resetting (no equivalent in TC);
- rolling stock has to guaranty some "EMC requirements", in particular regarding the electromagnetic field generated in direction of the detectors;
- the track checking periodicities have to be adapted to reduce the risk and the possible consequences of a BR;
- the track grinding periodicities[5] have to be adapted to reduce the risk and the possible consequences of a broken rail (preventive grinding 3 times a year ...).

[3] In France, a switch can authorize a new train for another direction 5s after the TC release

[4] In France, around 50% of the track circuits are jointless and this part is growing.

[5] It comes from the RIL 821 that: Rail wear during visual track supervision - visual (2 months); Internal Rail Integrity - Rail stability, rail cracks, rail welding stability, surface damage (4 months at 80km/h) and Rail Head Condition - (12 Months at 90km/h). In France, the periodicities of the internal rail integrity checking are divided by two if the track isn't equipped with track circuits.

From the standpoint of the railway system's design: Switching to AC would entail taking into account all the constraints stemming from the AC's inherent limitation. A deep study would need to be undertaken for each type of interlocking existing on each national network in order to assess the impacts on the system design on the one hand and on the overall balance sheet on the other hand. Such an economic study would involve comparing the expected benefits for the rolling stock with the costs tied to the redesign of the infrastructure (for example, an absolute block limits throughput) and the associated operating procedures (more frequent need of procedures to check that track sections have been cleared, for instance).

From the safety standpoint: The TC provides additional benefits that should not be underestimated, i.e.:

- Detection of broken rails, an advantage likely to grow in importance as concrete sleepers become more widespread on HSL (fast destruction of the sleepers);
- Detection of metal obstructions fallen onto the track, option to use the TC shorting bar or clip in emergencies or in maintenance (protection of a track section with workers) ...
- Phasing out TC would not respect the legal requirement to be "GALE" (Globally At Least Equivalent) in changing the system, from the standpoint of preventing the hazard of running over a BR at speed (function not fulfilled by AC). As a consequence, the loss of broken rail detection would require as-yet undefined compensating measures, e.g. close monitoring and grinding of rails with attendant higher maintenance, to be implemented in order to meet the GALE requirements, with a probabilistic point of view. Phasing out one for another system needs a new formal validation of the interlocking modules with the new type of detector (TC or AC).

4 Modelling

Estimation of the propagation of rail defects A global modelling of the two situations allows determining the pertinence domain (economic efficiency) while taking into account all the maintenance costs and the associated risk.

From a technical point of view, preventive grinding allows the elimination of early surface defects before they deteriorate thus extends the rail lifetime (augmentation of 30and has a positive effect on tamping durability.

From an economic point of view it remains to be determined whether the maintenance costs are reduced and the great investments related to rails replacement can be delayed.

From a safety point of view, preventive grinding doesn't allow the elimination of the internal defects and the possibility of a broken rail apparition has to be considered. Consequently, to estimate the residual risk with AC, we will take into account:

- the lifetime of rails based on a Weibull survival model for the remaining time before removal of rail sections;
- the velocity of decay of grindable rail defects;
- the impact of grinding on the survival function estimated before;

A Weibull survival model is used to model the rail removal function. The proba-bility for removing one meter section of rail is calculated. The cumulated distribution function of the Weibull distribution can be written as follows: $F(t) = 1 - exp(-(t/\eta)^{\beta})$ where β, η and $t > 0$. The variable t describes the age of the rail. The parameters β and η are estimated separately for every UIC group or HSL. The removals are considered as events; the rail sections without any removal are used as reference population and are right censored, i.e. there is no observed defect before the renewal of the line or before the end of the observation time. Thus, the date of their renewal or the date of the study is used as lower limit for the removal date. The parameters and are estimated by applying the least square method on the double logarithm of the empiric distribution function. The empiric distribution function is obtained by the Kaplan-Meier method, adapted for censored data. This method is used to estimate three distribution functions: the distribution of all defects, of grindable defects and of non-grindable defects. An example is given in Figure 1.

For each defect listed in the defect database, it is possible to find all control meas-urements that were carried out on this defect. Let P_i be the depth of a defect at the visit number i and t_i the corresponding date; the variable i then counts the number of visits carried out on the same defect. We calculate the increase of the defect between two visits, $\Delta P_i = P_i - P_{i-1}$. Several possible methods to estimate the propagation velocity were tested. The most intuitive estimator:

$$\hat{\upsilon} = \frac{1}{n} \sum_{observations} \frac{\Delta P_i}{\Delta t_i} \tag{1}$$

n is the number of all defect growths entered into the database and $\Delta t_i = t_i - t_{i-1}$, is not used, as it depends highly on the number of visits per defect.

The results were obtained with the estimator that is based on a linear regression for every defect. In general, the results obtained on the defect propagation concerns defects deeper than 5mm. The defects cannot be observed by ultrasonic inspections. The grinding interventions concern defects smaller than 1mm. We use an approximation as shown in Figure 2.

The observations in the defect database reveal that the most frequent defect grows under an angle of about 38° to the vertical. We estimate the velocity υ in the graph. Assuming the speed in the propagation direction remains constant, we have

$$\hat{\upsilon} = \upsilon \cdot \frac{sin(\alpha)}{cos(\beta)} \approx \upsilon \cdot 0,2204 \tag{2}$$

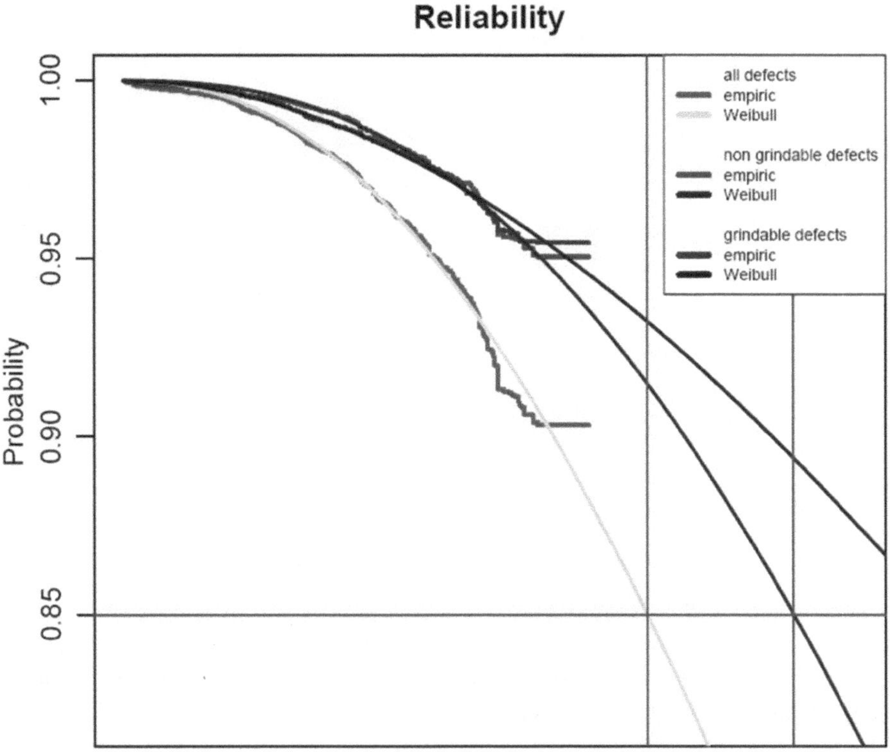

Fig. 1. Reliability of rails with respect to all defects, grindable defects and non grindable defects

The estimates of the propagation velocity will thus be multiplied by the factor 0.2204.

Impact of grinding on the reliability

Let's now identify the hypothetical curve corresponding to the distribution of all de-fects under preventive grinding from two reliability functions corresponding to grindable and non grindable defects. We use the following relation between the distribution functions: $R_{global(t)} = R_{grindable(t)} \cdot R_{nongrindable(t)}$ This relation is valid whenever a defect is either grindable or non grindable, and if the random variables describing the occurrence of this two defects are independent. It is supposed that grinding can extend the lifetime of the rail. A total renewal of the rail is needed if a given cumulated length has been replaced. Thus, as the end of life for the rail is defined by a fixed number of removals carried out, we can state that grinding "rejuvenates" the rail. This effect is used for the modelling, the formula for the virtual age function $fvirtualage(t)$ has the following form:
$fvirtualage(ti) = qgrindable((1-p) \cdot [Fgrindable(fvirtualage(t_i)) - Fgrindable(fvirtualage(t_i - 1))] + Fgrindable(fvirtualage(t_i - 1)))$

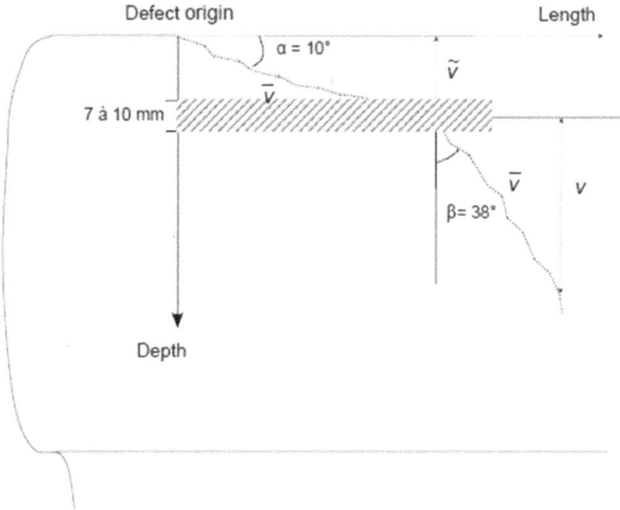

Fig. 2. Diagram of defect propagation

where $(1 - p)$ is the proportion of defects that remain after the grinding operation at time t_i;. To obtain the total distribution of the defects on a ground rail, we insert the distribution of grindable defects under cyclic grinding into the following formula: $Rglobal(t) = Rgrindable(fvirtualage(t)) \cdot Rnongrindable(t)$. We then obtain an approximation of the defect distribution for a ground rail even for combinations of grinding cycles and grinding depth that were never tested on real rails. An example of a $Rglobal(t)$ curve is given in Figure 3.

The study [2] shows this way that the optima determined by the cost minimizing algorithm are located between 2 and 4 year cycles and correspond to grinding depth of about 0.3mm depending on the UIC group (results valid on the conventional lines of the French network). The mathematical optima seem to reduce the rail maintenance costs down to 15

High speed lines A study realized on the HSL gives us the formula:

$$R_{global}(t) \approx [1 - e^{-(\frac{i}{\eta_1})^{1,45}}] \cdot [1 - e^{-(\frac{i}{\eta_2})^3}] \tag{3}$$

The parameters η_1 and η_2 are definite in regard of the cumulative traffic load. We can calculate the rails replacement rates on defect or rupture, by translating the grindable defects and those of internal tiredness of the rail (not grindable). In practice, it is necessary to take into account the effect of successive replacements [3] for the components in service. We define the function $h(t)$, the renewal density, that gives the number of replacements due to a failure. It can be shown that this function can be written as a sum of $n - convolutions$. There is no simple analytical expression for this function, even when considering the particular case of a Weibull distribution. The term can however be calculated numerically.

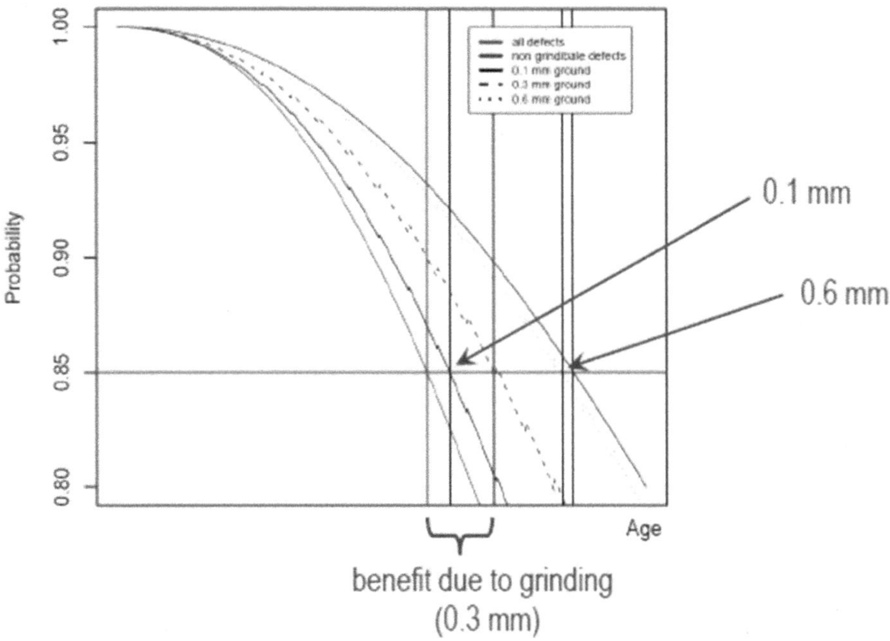

Fig. 3. Distribution of rail removals for a fixed grinding cycle depending on the grinding depth

The curve plotted on figures 4 shows that the two risks are at the same level. Thee internal rail defect can only been "neglected" in the first life time (around 10 years of use)!

The previous results confirm the data collected on the 3870km French high speed lines:

$$\lambda_{track} \approx 5 \cdot 10^{-6} / hour/trackkm \tag{4}$$

with an average of 16 year old rails.

Petri nets modelisation

The problem of comparing the economic and safety parameters of the two systems is not so simple. The Petri nets can be used to formalize the stochastical and logical links between the various situations.

The parameter x is the % rate of the undetected internal defect by the measuring cars, z is the % rate of non eliminated contact defect by the cyclic grinding operations and y is the % rate of detected broken rails by the track circuit. For HSL: $x \sim 4\%$, $z \sim 60\%$ (one grinding per year) and $y \sim 95\%$ (with out the switches).

For HSL: $x \sim 4\%$ (the detection by the driver comes to late), $z \sim 100\%$ (three grinding operations per year).

In the both cases, the simulation give us the distribution of the time to reach, either position 6, traducing the fact that the broken rail is crossed

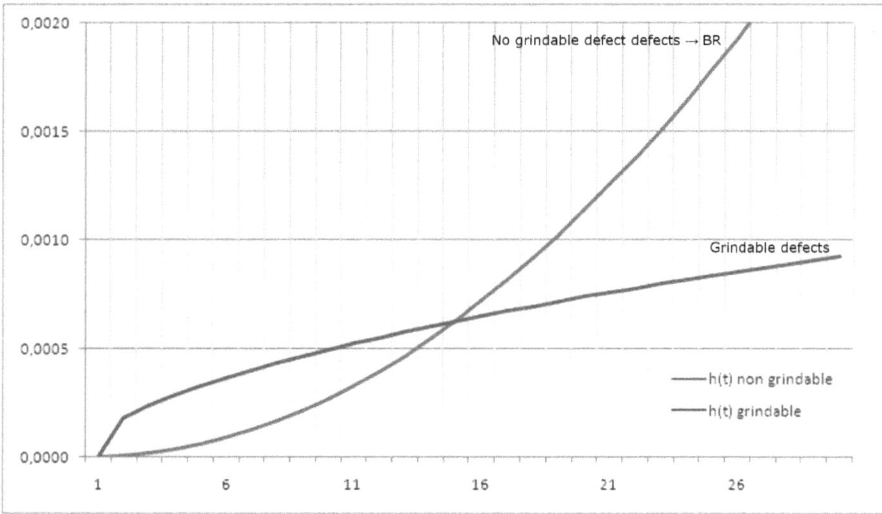

Fig. 4. The functions $h(t)$, the renewal density of grindable and non grindable rail defect

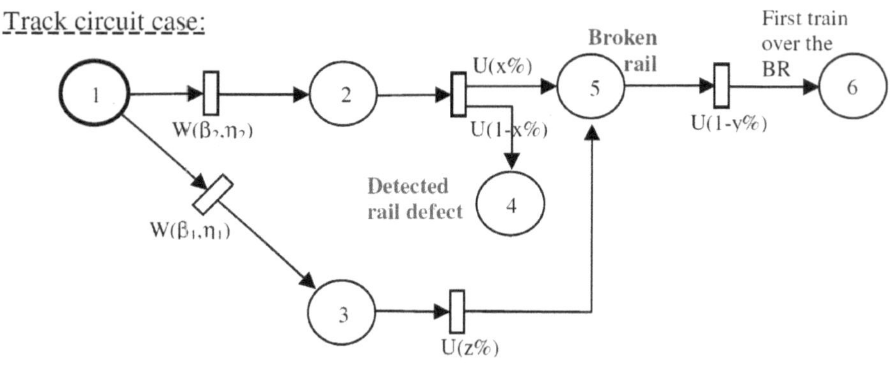

Fig. 5. The functions $h(t)$, the renewal density of grindable and non grindable rail defect

by a train and and, either position 4, traducing the occurrence on a rail replacement operation. The probability to over run a BR is 10 to 100 times higher with AC has with TC. We have to assume the risk of derailment on BK, the not respect of the legal requirement to be "GALE" (Globally At Least Equivalent), if we use AC alone. These results can be used to calculate the yearly costs taking into account all the factors, in particular for the rail and the signalling maintenance. In this case, other factors such as RAMS of equipment (e.g. transmitter/receiver of TC, axle counter itself) contribute to their cost.

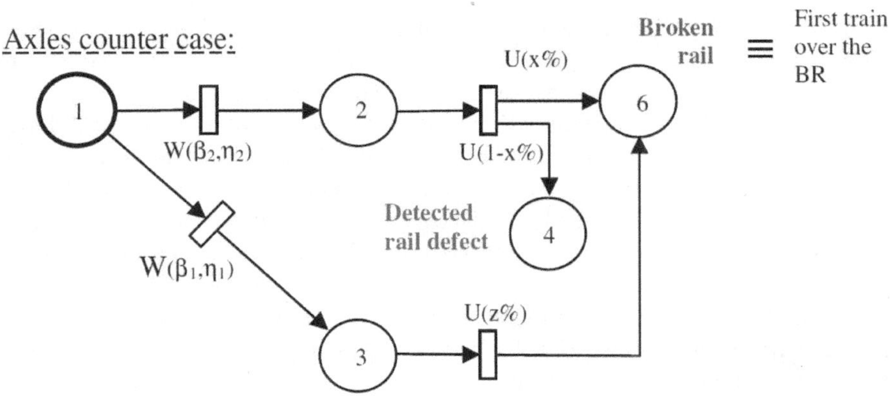

Fig. 6. The functions $h(t)$, the renewal density of grindable and non grindable rail defect

We assume that the rail defect and the associated derailment risk is a dominant factor for the choice of the track inoccupancy and track integrity detectors.

5 Conclusions

Each technology of TC or AC has its own domain of pertinence. Neither should be rejected out of hand. In the current state-of-the-art, neither of the two solutions allowing detecting the presence of trains is perfectly safe; signalling plant implicitly integrates the possibility of transient failures of the chosen solution.

The TC, through its occupancy detection function, provides, especially on high traffic or high speed lines, a higher level of safety because of lower risks from human inter-vention, as much in maintenance, particularly corrective maintenance, as in degraded mode operation. Moreover, it ensures a continual check of the electrical integrity of the track. The TC's reliability is all the better when the traffic is great and/or that the traction system is electric (typically 10^{-7}/h/TC on HSL). Moreover, the TC is well suited to traffic increase, since it allows the block to be permissive and sectional release in signal boxes. With the TC's, the safety level isn't coupled with the track maintenance policy, in particular regarding grinding.

The AC provides, in France, for low-traffic lines, an economical solution to the con-trol of clearing of long sections, without the risk of deshunting, albeit at the price of a stricter route locking (absolute block, no sectional release and so on) and greater operating complexity, in the event of disturbance or failure. The rate of failure internal rail defect rate is particularly low in this case.

The preventive grinding has been proven to minimise rail defects, to extend rail service life, to reduce the maintenance costs of the track but cannot give the guaranty of no broken rail (internal defect) on highly loaded tracks. Preventive grinding would require new technologies with high production output at low costs, as it is still large time slot consuming.

Safety: The broken rail detection issue will however remain, and will have to be solved, in particular for high speed lines. In general, the replacement from one type by an other type of detector, keeping all else constant, can lead to "deterministic" hazardous situations and has to be compensated. For example, the loss of broken rail detection would require as-yet undefined compensating measures, e.g. close monitoring and grinding of rails with attendant higher maintenance, to be implemented in order to meet the GALE requirements, with a "probabilistic" point of view. Phasing out one for another system needs a new formal validation of the existing interlocking modules with the new type of detector.

References

1. M. Antoni - P. Morand, "Migration of train detection systems for costs down for future rolling stock - Rolling Stock related view" - September 2009
2. C. Meier-Hirmer, Ph. Pouligny, M. Antoni "Impact of preventive grinding on maintenance costs and determination of an optimal grinding cycle" - SNCF, Infrastructure, Maintenance Engineering, Paris, France - ESREL 2008
3. M. Antoni, C. Meier-Hirmer "Economic correlation between maintenance and renewal - Optimization of maintenance strategies for tracks, signalling equipment and overhead line components" - SNCF, Infrastructure, Maintenance Engineering, Paris, France - WCRR 2008

Effects of a Periodic Maintenance on the Safety Integrity Level of a Control System *

Karol Rástočny and Juraj Ilavský

Department of Control and Information Systems, Faculty of Electrical Engineering, University of Zilina, Univerzitná, 010 26 Žilina, Slovak Republic
{karol.rastocny, juraj.ilavsky}@fel.uniza.sk

Abstract. An approach to analysis of impacts of preventive maintenance on safety integrity level of a control system is presented in the paper. Suggested approach incorporates combination of Continuous Time and Discrete Time Markov Chain which allows full employment of advantages of Continuous Time Markov Chains as a stochastic modelling method. Deterministic behaviour of either periodic or non-periodic preventive maintenance is consecutively implemented through Discrete Time Markov Chain.

Keywords: Safety, Periodic Maintenance, Safety Integrity, Control System, Markov Model.

1 Introduction

Many different quantitative methods can be employed if the safety integrity level (SIL) of a control system is to be analysed during a safety evaluation (even methods that were originally meant for reliability analyses - e.g. FMEA, RBD, FTA). Those widely used methods are supported by a large number of software tools that make possible not only a neat creation and editing of created models, but also provide a possibility of simulation and ways of presentation of achieved results [1].

The aim of a qualitative evaluation of the SIL of a system is to determine probability of a hazardous state of a system (probability that a specific safety function fails) that depends on many safety-influencing factors [2], [3], i.e.:

$$p_H(t) = f(\lambda, c, \mu, t_D, \ldots, t) \tag{1}$$

Given the fact that the SIL of a system rests on many miscellaneous factors (random hardware failure rate λ, diagnostic coverage coefficient c, time needed to detect a failure t_D, recovery rate μ, ...), quantitative method for the SIL evaluation that covers most of the factors is usually preferred. Such

* This paper was supported by the scientific grant agency VEGA, grant No. VEGA-1/0040/08 "Mathematic-graphical modelling of safety attributes of safety-critical control systems".

E. Schnieder, G. Tarnai (eds.), *FORMS/FORMAT 2010*,
DOI 10.1007/978-3-642-14261-1_8,

a method should also be rather comprehensive and undemanding in the terms of computational power [4], [5]. Markov chains meet all these requirements.

Paper is focused on problems that arise when an impact of a recovery on the SIL is quantitatively analysed. There have already been attempts to quantify the impacts of the recovery on the safety of the system [2], based on either numerical solving methods or ones that derived a closed-form equations valid under rather restrictive assumptions. Presented solution of those problems is based on a convenient combination of the Continuous Time Markov Chains (CTMC) and the Discrete Time Markov Chains (DTMC).

2 Theoretical basis

Principals of the Markov model of the impacts of a recovery on the SIL are shown in the Fig. 1. Creation of the Markov model can be divided into two phases.

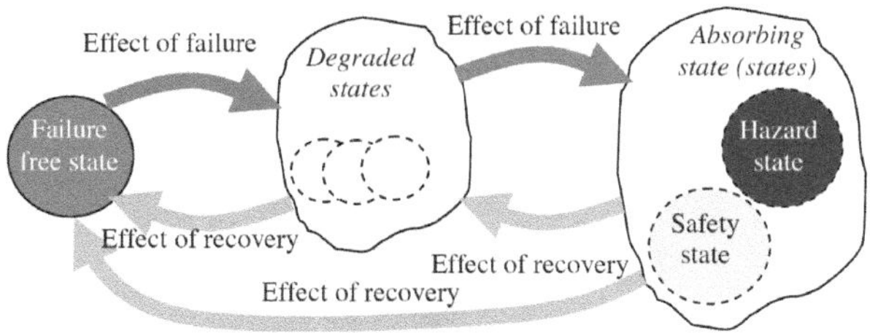

Fig. 1. Important states of a control system in SIL analysis

Identification of stochastic events and their outcomes (hardware failures and failure modes for instance) is the first phase of a Markov analysis. Identified events have in common that their occurrence makes the system change its state from the initial state F (failure free, safe state), to several degraded, but still operating and safe states. System eventually ends in an absorbing state. One absorbing state in the model, the hazardous state H, is always necessary. Optionally, there can be two absorbing states in the Markov model in safety analysis; the other one would be the safe state S (that is reached after a detection and negation of a detectable failure that occurred in the system). If the time instants of events occurring in the system are an exponentially distributed random variable, then Markov chain describing such a system is referred to as homogenous CTMC. Transition rates of the homogenous CTMC are constant, which means that such a model is invariant in shifts in time (an analogy to the exponential distribution).

$$P(t < T \leq t + t_0 | T > t) = P(T \leq t_0) \tag{2}$$

This property of the homogenous CTMCs simplifies the control system safety analysis process and created models are usually easy to analyse via analytical methods.

In the second phase of Markov analysis changes in system caused by recovery mechanisms (full or partial system recovery) are assessed. Following situations are relevant to modelling when recovery of the system is taken into account:

- Partial recovery of the system after a part of the system had failed and failure has been detected by a detection mechanism, but system remained in the operational state. This could happen in multi-channel redundant system that allows required function to be performed even after a partial failure has occurred (2-out-of-3 system for instance). The part of the system that had failed is restored only after a fault correction and function check-out has been completed.
- Recovery of the system after a failure from a safe state (down state) - when the system after the detection and negation of the failure reached the safe state that could be abandoned only after a fault correction and function check-out of the entire system has been completed.
- System restoration after a preventive maintenance. If an online diagnosis is not able to detect all possible hazardous failures ($c < 1$), then preventive maintenance has to be performed on a periodic basis. Periodic preventive maintenance is fo-cused on failures that remained undetected during operation of the system.

However, it is a valid assumption that even after the system recovery (partial or full) some undetected failures may still be present in the system [2], [6].

Homogenous CTMC is entirely described by an infinitesimal generator matrix A and a row vector of initial probability distribution

$$\overrightarrow{p(0)} = \{p_1(0), p_2(0), \ldots, p_n(0)\}, \tag{3}$$

where n is total number of states in the model.

If we equate $p_1(t)$ with $p_F(t)$ (the probability of a failure-free state), then

$$\overrightarrow{p(0)} = \{1, 0, \ldots, 0\}. \tag{4}$$

By solving the linear differential equation system

$$\frac{d}{dt}\overrightarrow{p(t)} = \overrightarrow{p(t)} \cdot \mathbf{A}, \tag{5}$$

for every initial probability distribution the time-dependent probability distribution can be determined

$$\overrightarrow{p(t)} = \{p_1(t), p_2(t), \ldots, p_n(t)\}. \tag{6}$$

While CTMC describes the stochastic occurrence of the events in the system, DTMC allows us to encompass deterministic periodic preventive maintenance in the same model.

In general it is believed that the preventive maintenance has an 100% diagnostic coverage and is able to detect all failures that are present in that time in the system. Immediately after the preventive maintenance is completed the system is with probability p=1 restored to initial, failure-free state. From this assumption we can for perfect preventive maintenance establish the DTMC transition probability matrix

$$\mathbf{P}_{nxn} = \begin{pmatrix} 1 & 0 & \cdots & 0 \\ 1 & 0 & \cdots & 0 \\ \vdots & \vdots & \ddots & \vdots \\ 1 & 0 & \cdots & 0 \end{pmatrix} \tag{7}$$

Generalised transition probability matrix for any diagnostic coverage of the preventive maintenance can be declared as

$$\mathbf{P}_{nxn} = \begin{pmatrix} \overrightarrow{\pi_1} \\ \overrightarrow{\pi_2} \\ \vdots \\ \overrightarrow{\pi_n} \end{pmatrix}, \tag{8}$$

where $\overrightarrow{\pi_i}$ is a row vector with size n. The quantities π_{ij} fulfil the following condition:

$$\sum_{j=1}^{n} \pi_{ij} = 1. \tag{9}$$

After the preventive maintenance has been performed and function checkout has been carried out, system can be restored back into operation. That can be modelled, with respect to (2) as a return of the simulation to the time $t = 0$, with corresponding initial probability distribution. New initial probability distribution is determined by multiplication of time-dependent probability distribution in the system restoration time t and transition probability matrix $\mathbf{P}$, i.e.

$$\overrightarrow{p^k(0)} = \overrightarrow{p^{k-1}(t)} \cdot \mathbf{P}, \tag{10}$$

where k subscript refers to the operation phase after the k-th restoration of the system, $k \geq 1$ and the vector $p^0(0)$ is determined by (4).

Given the fact that CTMC remains constant after every recovery of the system, it is convenient to find a general solution for the linear differential equation system (5). Then a particular solution for every operation phase can be derived by setting the initial probability distribution for corresponding phase after every restoration of the system (10). Note that this solution also allows us to consider and analyse effects of non-periodic preventive maintenance. If the failure rates of the components could rise over time, then approximation could be made via modified constant failure rate in every consecutive phase.

3 Case study

Let us assume safety relevant control system with 2-out-of-2 system architecture. Such a system is depicted in the Fig. 2 via a simplified block diagram.

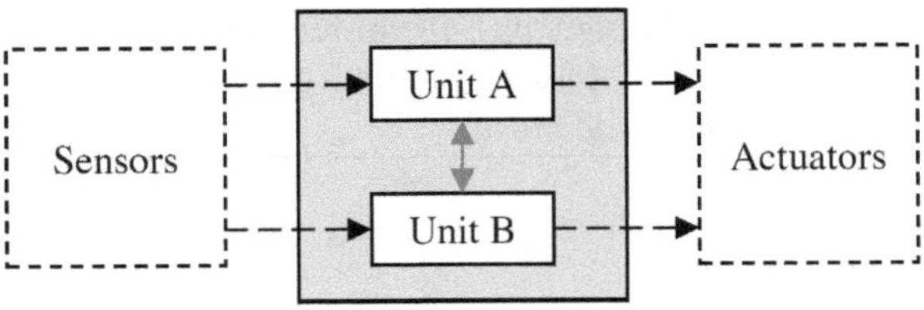

Fig. 2. Block diagram of a general 2oo2 control system

Further assumptions for this case study are: only units A and B directly influence the safety of the system; both A and B units are identical, therefore hardware failure rates are $\lambda_A = \lambda_B = \lambda$; on-line diagnosis with its detection mechanisms has the diagnostic coverage coefficient $c < 1$ and the system is able to reach the safe state after a detectable failure has occurred. Transition rate to the safe state δ is determined through a time needed to detect a failure and negate its consequences.

If all conditions stated above apply, then impacts of the hardware failures on the SIL could be analysed by the CTMC shown in Fig. 3. Further meanings of states and transitions is summarised in the Table 1 and Table 2.

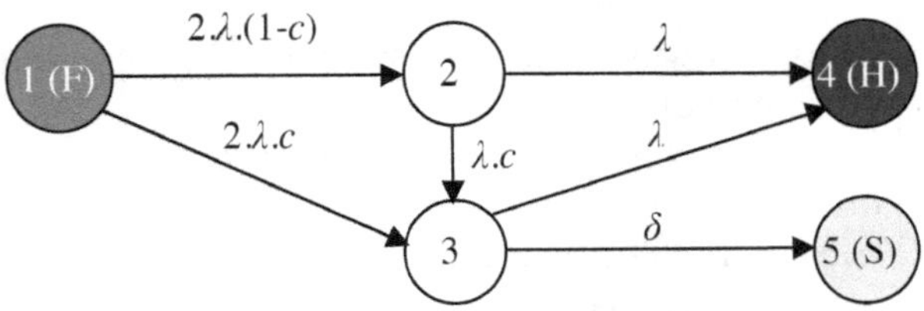

Fig. 3. CTMC of a general 2oo2 control system

CTMC in the Fig. 3 is determined through the infinitesimal generator matrix

Table 1. Description of the states of the diagram in Fig. 3

State	Description
1	Initial, failure-free state in the time $t = 0$ (safe state)
2	Random *undetectable* failure is present in either the A or B unit (safe state)
3	Random *detectable* failure is present in either the A or B unit (safe state)
4	A failure is present in both units (hazardous state)
5	The state after a failure of any of the units has been detected and negated (safe state)

Table 2. Description of the transitions of the diagram in Fig. 3

Transition	Description
$1 \rightarrow 2$	Occurrence of a random *undetectable* failure of either A or B unit
$1 \rightarrow 3$	Occurrence of a random *detectable* failure of either A or B unit
$2 \rightarrow 3$	Occurrence of a random *detectable* failure of the unit that already has an undetectable failure
$2 \rightarrow 4$	Transition will occur when the second (operational) unit fails after any failure of the first unit
$3 \rightarrow 4$	Transition will occur when the second unit (operational) fails after any failure of the first unit
$3 \rightarrow 5$	Transition will occur when the detectable failure of either A or B unit has been detected and negated

$$\mathbf{A} = \begin{pmatrix} -2\lambda & 2\lambda(1-c) & 2\lambda c & 0 & 0 \\ 0 & -\lambda(1-c) & \lambda c & \lambda & 0 \\ 0 & 0 & -\lambda-\delta & \lambda & \delta \\ 0 & 0 & 0 & 0 & 0 \\ 0 & 0 & 0 & 0 & 0 \end{pmatrix} \tag{11}$$

Given differential equations system (5), infinitesimal generator matrix (11) and initial probability distribution (4) time-dependent probability of the hazardous state of the system $p_H(t)$ can be calculated. $p_H(t)$ function is plotted in the Fig. 5 (dashed line). Preventive maintenance does not change the state-space of the system; therefore DTMC model (Fig. 4.) has the same number of states with the same meaning as the CTMC model.

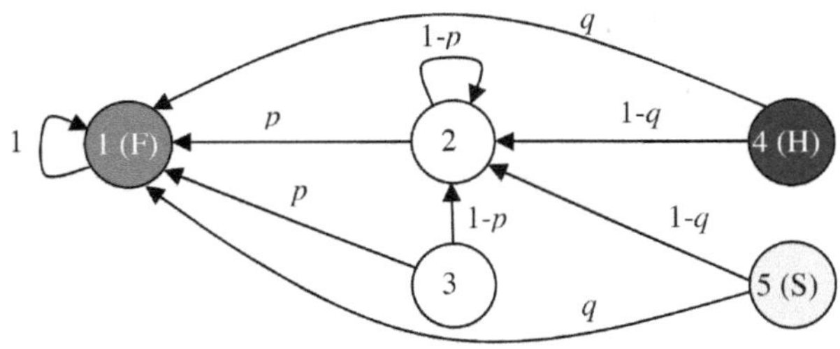

Fig. 4. DTMC model of a general 2-out-of-2 control system

Diagram in the Fig. 4 shows states of the system and transitions with corresponding transition probabilities. It is clear that after a system restoration the system will be either in the 1st state (if all random hardware failures has been detected and repaired during the preventive maintenance) or in the 2nd state (if not all random hardware failures has been detected and repaired during the preventive maintenance). Probability of the system being in the 1st or 2nd state after the restoration is determined by probability transition matrix (12).

Various maintenance operations or sequences of operations could be executed depending on the state where the system would be at the instant when preventive maintenance starts. This fact was considered and implemented in the model via an unequal probabilities p, q in the diagram in the Fig. 4. However, it should be noted that for high safety integrity level systems it is very unlikely that the value of q would be lesser than 1.

$$\mathbf{P} = \begin{pmatrix} 1 & 0 & 0 & 0 & 0 \\ p & (1-p) & 0 & 0 & 0 \\ p & (1-p) & 0 & 0 & 0 \\ q & (1-q) & 0 & 0 & 0 \\ q & (1-q) & 0 & 0 & 0 \end{pmatrix} \tag{12}$$

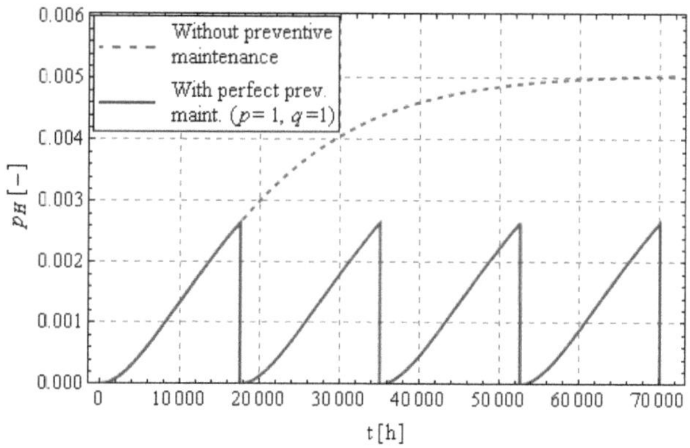

Fig. 5. Probability of the hazardous state for the 2-out-of-2 system with $\lambda = 5 \cdot 10^{-5}h^{-1}$, $c = 0,99$, $\delta = 1h^{-1}$

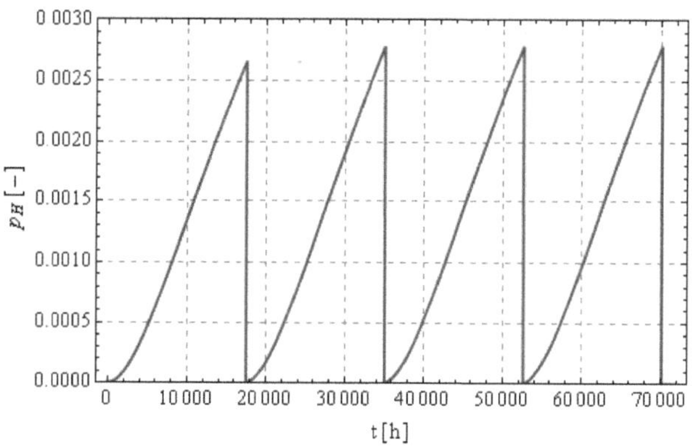

Fig. 6. Probability of the hazardous state for the 2-out-of-2 system with $\lambda = 5 \cdot 10^{-5}h^{-1}$, $c = 0,99$, $\delta = 1h^{-1}$, $p = 0,9$ and $q = 1$

Fig. 5 compares the $p_H(t)$ of the not preventively maintained system to the system which is preventively maintained every other year and whose ideal preventive maintenance is capable of detecting all failures that can occur in the system. Plot in the Fig. 5 shows that in the case of perfect preventive maintenance the maximum value of the $p_H(t)$ during system lifespan depends only on the duration between preventive maintenances and does not depend on the total up time of the system.

In contrast to Fig. 5 the plot in the Fig. 6 shows that maximum value of $p_H(t)$ of the system with imperfect preventive maintenance increases with every consecutive phase of operation. Since the system is down during preventive maintenance, therefore it can be assumed that $p_H(t)$ is independent of the duration of the preventive maintenance, even though it cannot be directly inferred from the Fig. 5.

Increments in the maximum value of the $p_H(t)$ in every consecutive phase are directly caused by a poor preventive maintenance ($p = 0,9$) that cannot detect all system failures. Increments would be even more severe if we assumed $q < 1$, but that case is in high SIL safety-critical systems avoided at all costs.

4 Conclusion

System restoration could possibly happen after an unscheduled maintenance that is usually required after the system has reached non-transient safe state (absorbing state). Unscheduled maintenance shortens the deterministic period between two consecutive preventive maintenances. However, this fact has no negative effect on computed probability of the hazardous state as that will always be higher than the probability of the hazardous state in the instant of unscheduled maintenance. This is in full accordance with the safety analysis philosophy.

Such analysis approach has been successfully employed to assess the safety of the electronic interlocking systems used in ŽSR operation, even though those systems are generally more complex and span over more levels of control.

References

1. Rástočný, K.: Modelling of Failure Effects to Integrity of System. In international scientific journal: Advances in Electrical and Electronic Engineering, No. 2 Vol. 3, pp. 91-94, ISSN 1336-1376, (2004)
2. Brissaud, F., Barros, A., Berenguer, C.: Probability of failure of safety-critical systems subject to partial tests. Reliability and Maintainability Symposium (RAMS) 2010 Proceedings - Annual, pp. 1-6, Digital Object Identifier: 10.1109/RAMS.2010.5447972, (2010)
3. Rástočný, K., Foltan, M.: Primary factors Affecting Safety of Control System. In international scientific journal: Advances in Electrical and Electronic Engineering, No. 1-2 Vol. 5, pp. 154 -157, ISSN 1336-1376, (2006)
4. Klapka, Š., Súkup, J.: New approach to the automated creation of Markov Chains for evaluating dependability parameters of complex redundant systems. In: FORMS/FORMAT 2004, pp. 24-29, Braunschweig, Germany, ISBN 3-9803363-8-7, (2004)
5. Rástočný, K., Janota, A., Zahradník, J.: Problems Related to Quantitative Appraisal of Safety-Related Systems. Proceeding of the international scientific symposium FORMS/FORMAT 2007, Braunschweig, Germany, pp. 319-325, ISBN 13:978-3-937655-09-3, (2007)
6. Bukowski, J. V.: Modeling and analyzing the effects of periodic inspection on the performance of safety-critical systems. IEEE Transactions on Reliability. Volume: 50, Issue: 3, pp. 321 - 329, Digital Object Identifier: 10.1109/24.974130, (2001)

Modeling Computer based, microscopic Dispatching Systems

Alexander Kuckelberg[1,2] and Ekkehard Wendler[2]

[1] VIA Consulting & Development GmbH, Aachen, Germany
a.kuckelberg@via-con.de,
http://www.via-con.de
[2] VIA - Institute of Transport Science, Aachen, Germany
{wendler/kuckelberg}@via.rwth-aachen.de,
http://www.via.rwth-aachen.de

Abstract. Computer aided dispatching systems for railway operations are designed to support dispatchers when detecting arising problems or undesirable situations within daily operation.
On a microscopic level such systems can even be designed to determine detailed routing information, time margins and dwell times in such a way, that occupation times implied by train control systems can be calculated. This assures, that the dispatching decisions remain valid with respect to operability.
While the core component, the dispatching algorithms and approaches are common research topics an overall consideration of system modeling and integration within existing systems is less commonly evaluated. This paper tries to introduce a generic approach to dispatching system modeling by analyzing fundamental functionalities and by abstract system definitions.

Keywords: Dispatching Systems, Microscopic, Framework, System Architecture

1 Introduction

The operation of computer aided dispatching is a quite special application field whose importance and acceptance is increasingly growing with the implementation of computer based train control and communication technologies. While the implementation of such control and communication technologies is a great leap forward to efficient train operation the next steps may be the extensions of these systems to support train dispatchers and operators.

Assuming that a sufficient train operation control technology is available, dispatching systems are usually added and integrated into these existing systems. With a published timetable as a reference point the main functionalities of a dispatching system are the detection of train movements and positions, the alignment of these positions to the intended timetable, the detection and

E. Schnieder, G. Tarnai (eds.), *FORMS/FORMAT 2010*,
DOI 10.1007/978-3-642-14261-1_9,

prognosis of problems and undesired situations, the determination of sufficient dispatching actions and finally the propagation and realization of these actions. At the end, the realization of actions is monitored by train movement detection again. This process cycle is shown in Figure 1, which originally was introduced in [1] to illustrate the bordering conditions within which an asynchronous dispatching algorithm was evaluated.

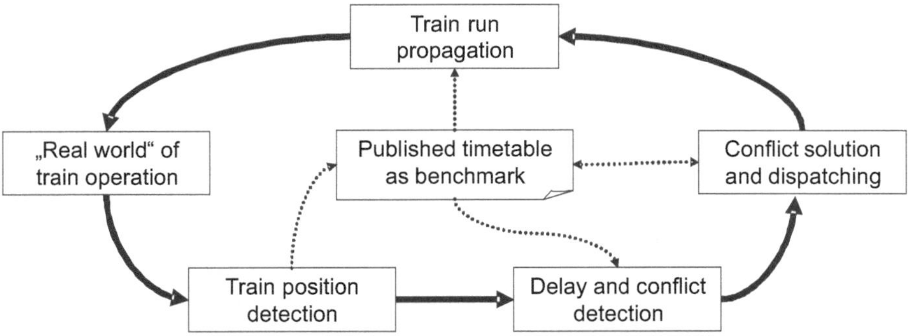

Fig. 1. Integration and functionalities of computer based dispatching systems.

The different functionalities and components can be introduced shortly:

- **Published timetable and basic data structures** The timetable is the central element for reliable, efficient and predictable train operation and the core data structure for a dispatching system: The system starts with the published timetable $tt_{orig} = dp_0$. The forecasted train runs are based on this timetable, which is modified successively due to detected train movements, their reported positions and dispatching decisions made. Each modification transforms dp_i into another timetable dp_{i+1}, often called the *dispatching plan*. All operations are always applied to the last valid dispatching plan.
- **Train position detection** A correct and reliable detection of train positions and the derivation of train movements is a requisite for dispatching systems. Different solutions are possible to enable discrete or continuous detection. The detection based on discrete position messages may be the most common approach, because systems with discrete approaches are widely available and standardized position telegrams are specified by the UIC [2].
These telegrams encapsulate a time t, a train tr and a position p. With this information the route r_{orig} of tr is derived from the current dispatching plan dp_i. If p is found within r_{orig} the train trajectory is adjusted in p at t. Otherwise a new route r_{new} containing p must be determined and a new trajectory is calculated. The new route *could* be the current one, the corresponding trajectory again is adjusted in p at t. A successful adjustment transforms dp_i into dp_{i+1} containing the adjusted trajectory.

- **Delay and conflict detection** Within a dispatching plan arising disturbances (e.g. a high train density) and conflicts (e.g. overlapping blocking times) have to be detected. The conflict time detection uses reduction functions to simulate the uncertainty of future operation situations, conflicts beyond a time horizon can be ignored.
- **Rescheduling** If conflicts are detected within a dispatching plan dp_i, a new conflict free plan dp_{i+1} is produces by conflict solving components. The primary goal of a microscopic dispatching system is the elimination of blocking time overlaps. If equivalent solutions exist, additional weighting criteria are applied, e.g. weighted sum of new delays, kept train links etc. Several algorithms and approaches for conflict solving algorithms exists, a good overview and evaluation is given e.g. in [3].
- **Dispatching plan propagation** Decisions made by dispatching systems must be propagated and submitted to interlocking stations and/or train drivers. Depending on the technical, organizational and legal environment different integration levels are possible. The propagation process can be a safety relevant task of such systems.

With this rough introduction of computer based dispatching systems and there integration within the operational environment the next chapter analyses the functional components of such systems more detailed and introduces interactions and data flows.

2 System Architecture

Following the rough overview of computer based dispatching systems and their tasks, a generic system architecture can be set up. The architecture must contain components covering the required functionalities:

- Basic infrastructure and timetable data must be provided when the system is initialized. The initialization must provide microscopic infrastructure data for routing algorithms of the dispatching system and for detailed run time and occupation computation as well as timetable data containing the planed trajectories, allocated travel times, departure and arrival times for delay evaluation and stopping policies to evaluate valid routes.
- Communication channels provide dispatching systems with required real time information. As evaluated e.g. by the DisKon-project [4] two kinds of system messages are seriously evaluable by computer based dispatching systems: messages associated to a specific train and messages containing interlocking data.

  Train related messages are e.g. messages about train positions, delay reports or notifications about the change of train characteristics and properties. Interlocking messages reflect information about routes currently set or released by interlocking stations or block occupations. These messages are finally evaluated and the derived information is integrated within the current timetable, possibly leading to modified train trajectories.

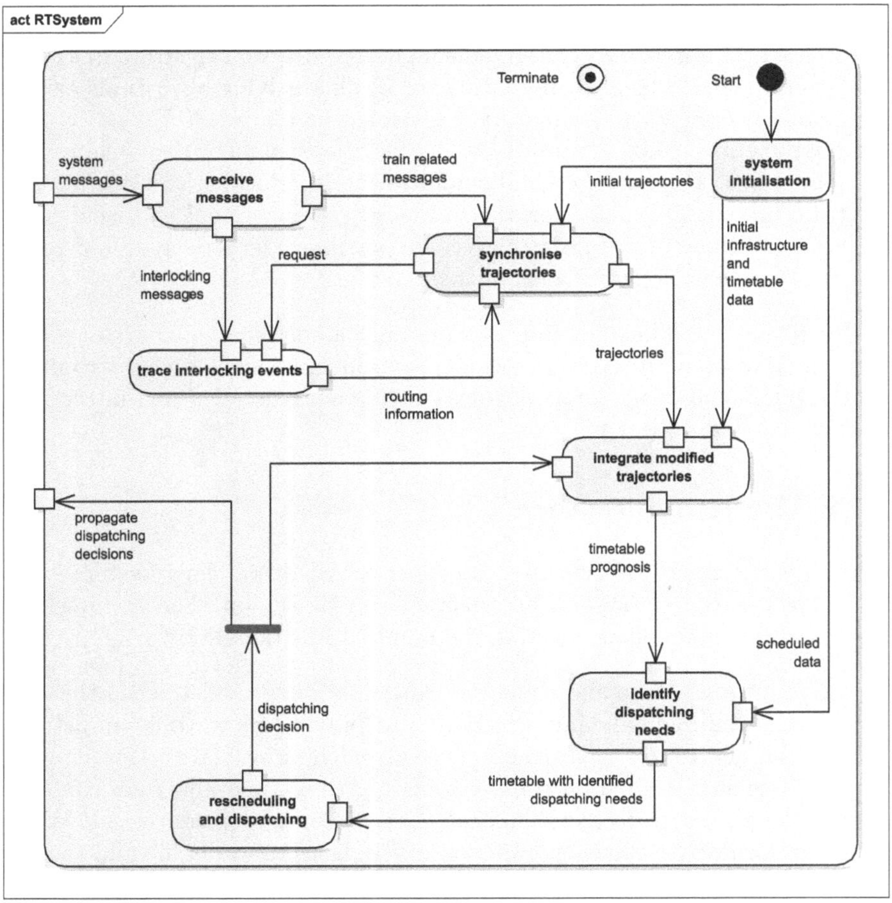

Fig. 2. System activities and interactions.

- Delay and conflict detection requires the evaluation of the current timetable (containing all adjusted trajectories) with respect to occupation overlaps within defined tolerance and time horizon ranges, the detection of train connection failure risks, the analysis of possible employee or train vehicle circulation etc. Whatever is evaluated and considered as a problem respectively a conflict strongly depends on the real implementation and on the kind of information available. Moreover the dispatching algorithm also influences the implementation of the conflict detection component.
- The conflict solution component has to solve the identified conflicts and problems. The algorithm realizing this solving may vary and follow different approaches like synchronous or asynchronous paradigms, rule based, functional or heuristic approaches and much more. As mentioned before the realization of the dispatching algorithm is the most common subject of research activities in the context of dispatching systems, including theoretical consideration with respect to optimality of found dispatching decisions, the performance, complexity or scalability.
- Dispatching decisions must be integrated within the prognosis on the one hand and - to make the prognosis probable - propagated into the real operation on the other hand. The propagation again is varying and depending on the practical conditions under which the system is working and the grade of integration within existing systems. A very strong integration would be the direct influence of operation systems, the direct transmission to train drivers or engines, the information of train dispatchers, service staff or passenger information systems. A weak integration would be a simple display for a few people, which have to decide, what to do with the displayed dispatching system decisions.

Figure 2 shows the different activities and their interactions, which have to be covered by components of the system architecture. The activities can be performed by single components, the major interaction and communication flow between the activities correspond to the once of the components. The parallelism and scalability of dispatching system components can be derived from the parallelism of the activities. In the following, activities are often denoted as (system) components implementing them in a 1:1-relationship.

Components receiving messages can be set up for each communication channel delivering messages. These components are independent from each other and act like buffers between operation and the dispatching system. The components also uncouples the operation time (real time) from the dispatching system time. It can deliver messages to system internal components on request, not on a predefined time behavior.

Components processing system messages can be scheduled in parallel according to the kind of message. In principle each train and its trajectory can be handled by a component, train related messages can be assigned to and evaluated by the component associated to the train. To enrich trajectory information interlocking information can be requested from components handling interlocking messages. If routing information are available, e.g. allocated routes, these information might be integrated within the trajectories to support dispatching components to find good solutions. Delay detection can be performed by these components as well, because delays are considered train wise.

With respect to parallelism the detection of conflicts is an algorithmic bottleneck, because the conflict detection has to synchronies all train trajectories to detect occupation overlaps. Also the detection of vehicle circulation, connecting train links etc. requires the association of several trajectories. Therefore parallel conflict detection requires advanced and complex processing and synchronization mechanisms, but nevertheless with approaches like regional division of dispatching areas [5] the conflict detection can be scheduled in parallel, too.

3 Modeling Dispatching Systems

Starting from the introduced system architecture point of view, this chapter presents a formal specification for (microscopic) dispatching systems. Referring to the introduction, a published timetable $tt_{orig} = dp_0$ must be available as a reference point for dispatching. Within the dispatching and rescheduling operation dp_i is transformed into another plan dp_{i+1}, called the dispatching plan. The dispatching plan with the greatest index is the current and valid one. The domain of an object o is denoted as $Dom(o)$

Definition 1 ((dispatching system)). *Let execute : void → void be a function without parameters and return values. Then*

$$DS = (CDs, CSs, dp, PPs, < execute >)$$

is a dispatching system with a non-empty set $CDs = \{cd_1, \cdots, cd_m\}$ $(m \geq 1)$ *of conflict detection objects (definition 2),*
a non-empty set $CSs = \{cs_1, \cdots, cs_n\}$ $(n \geq 1)$ *of conflict solving objects (definition 3), a dispatching plan dp, a probably empty set* $PPs = \{pp_1, \cdots, pp_p\}$ $(p \geq 0)$ *of dispatching propagation objects and the execute-method.*

The *conflict detection*-objects implement the conflict detection functionality applied to *dp*. These objects return a set of detected conflicts, which have to be solved by *conflict solving*-objects, which perform the rescheduling and dispatching activities. These conflict solving objects return a set of dispatching decisions, which are propagated through *dispatching*-objects and applied to the current dispatching plan.

The *execute*-method performs the cyclic train detection and dispatching activities performing the internal dispatching process cycle:

while not terminated do {

(1) $\forall_{cd \in CDs, c=\emptyset} \; c = c \cup cd.getConflicts(dp);$

(2) $\forall_{cs \in CSs, d=\emptyset} \; d = d \cup cs.getDispatchingDecision(c);$

(3) $\forall_{pp \in PPs} pp.propagateDecision(d);$

(4) $dp.integrateDecision(d);$

}

This cycle implements the basic steps of conflict detection (1), conflict solving (2) and dispatching decision propagation (3+4). Components for these tasks are specified in the following.

Definition 2 ((conflict detection component)). *Let dp be a dispatching plan, t the published timetable and $\mathcal{C}$ the domain of (microscopic) conflicts. Then a conflict decision component CD is a tuple*

$$CD = (t, dp, < getConflicts >)$$

where $getConflicts : Dom(dp) \to \mathcal{C}$ is a method detecting all conflicts of the current dispatching plan with respect to the reference timetable t.

Objects of this class perform the evaluation of dispatching plans and the detection of problems and undesired situations arising due to the train protection based prognosis. The dispatching plan used for conflict detection should be practically speaking the current one. The rescheduling or conflict solving activity is performed by conflict solving components.

Definition 3 ((conflict solving component)). *Let dp be a dispatching plan and $\mathcal{D}$ the domain of dispatching decisions. Then a conflict solving component CS is a tuple*

$$CS = (dp, < getDispatchingDecision >)$$

where $getDispatchingDecision : Dom(dp) \times \mathcal{C} \to \mathcal{D}$ is a method, which determine dispatching decisions for provided conflicts and a specific dispatching plan.

With the three preceding definitions a quite abstract modeling of dispatching systems (not even restricted to computer based systems) is introduced. The major intention of this abstract approach is the separation of functionality and concrete implementation aspects. The approach is open for different and varying realization alternatives with respect to detected conflicts as well as to algorithms and approaches chosen for modifying dispatching plans.

Finally the missing functionality, the propagation of dispatching decision, is specified in a similar generic way. While the propagation targeting the real operation strongly depend on framework conditions of existing control systems, the dispatching plan of the dispatching system has to consider the dispatching decision in any way.

Definition 4. *(dispatching propagation component) A dispatching propagation component PP is a tuple*

$$PP = (< propagateDecision >)$$

where $propagateDecision : \mathcal{D} \rightarrow void$ *is a method, which receives a dispatching decision and propagates it into train operation.*

Finally the last specification introduced here is the dispatching plan, which is used by conflict detection and conflict solving components and when integrating the dispatching decision.

Definition 5. *(dispatching plan) A dispatching plan* dp *is a tuple*

$$dp = (TJs, < integrateDecision >)$$

where TJs *is a set of train trajectories contained within the plan and a method* $integrateDecision : \mathcal{D} \rightarrow void$*, which integrates dispatching decisions into the plan.*

Several additional components represented within the system architecture have to be disregarded, but similar specifications can be set up for components e.g. managing operation messages.

4 Conclusion

Inspired by the question, how computer aided dispatching systems for railway operations should be designed; this paper presents an approach and abstract specification of dispatching decision systems. All considerations are based on microscopic models.

This paper first introduced the dispatching systems and their system architecture in a quite abstract way before specifying some activities and components formally and in a generic and universal manner, which abstracts from concrete implementations.

In this way, this paper evaluates and identifies the fundamental functionalities of dispatching systems and transforms this knowledge into a model, which might be used as a framework for dispatching system specification, description and implementation. Moreover the specifications provide a basis for further research and modeling activities.

Finally it can be stated, that systems following the component based system architecture like the prototype realized within the DisKon-project [4] have shown, that the approach chosen and presented in this paper is functional and applicable for real time dispatching systems.

References

1. Kuckelberg, A., Wendler, E.: Real-time Asynchronous Conflict Solving Algorithms for Computer Aided Train Dispatching Assistance Systems. In: Computers in Railways (Comprail) XI (2008), S. 555 - 563.
2. Pause, T.: Systeme für Disposition und Lokalisierung, EI-Eisenbahningenieur, 05/08, pp. 21-24.

3. D'Ariano, Andrea: Improving Real-Time Train Dispatching: Models, Algorithms and Applications. 2008, Delft University of Technology, The Netherlands.
4. DisKon - Rat aus dem Rechner, bahntech 02/07, pp. 4-9, 2007, Deutsche Bahn AG, www.db.de/bahntech
5. Kuckelberg, A., Kurby, S.: Kopplung mikroskopischer Belegungsdisposition und makroskopischer Anschlussdisposition im Eisenbahnnetz., Proceedings of 21. Verkehrswissenschaftliche Tage, Dresden, 2007

A Method of Evaluating Railway Signalling System Based on RAMS Concept

Shigeto Hiraguri, Koji Iwata, and Ikuo Watanabe

Railway Technical Research Institute,
2-8-38, Hikari-cho, Kokubunji-city, Tokyo, Japan
{hiraguri, kouji-i, ikuo}@rtri.or.jp

Abstract. Since the IEC 62278 (RAMS Standard) was published in 2002, its importance and impact to Japan have been conspicuous. On the contrary, various new train control systems based on information technology are under active development. Considering such situations, we assume that it is essential to establish an evaluation method to design railway signalling systems which have an excellent balance to the RAMS indicators at low costs. We propose a basic method to evaluate the performance of a signalling system by a cost indicator based on the concept of RAMS, and confirmed that the method can evaluate systems appropriately considering grade of the line, circumstances and other factors.

Keywords: Railway Signalling, RAMS, Cost, System Architecture, Evaluation

1 Introduction

A railway signalling system has been contributing to ensure safety of railway transport for long years. However, conventional systems were developed gradually by bottom-up or experience-based approaches. In 2002, IEC 62278 (RAMS standard) of which a basic concept is to realise excellent performance by a systematic approach. Furthermore, various new train control systems based on information technology are under active development. Considering these situations, it is required to establish an evaluation method which is applicable to design the signalling system, which has an excellent RAMS performance and appropriate functional architecture adapting to a given circumstance.

2 Railway Signalling System and RAMS Indicators

There are several types of conventional signalling systems, according to characteristics such as the grade of the line. When a railway company introduces a new signalling system, experts comprehensively consider the grade of the

E. Schnieder, G. Tarnai (eds.), *FORMS/FORMAT 2010*,
DOI 10.1007/978-3-642-14261-1_10, © Springer-Verlag Berlin Heidelberg 2011

line, the amount of transport, the track layout in the station, the initial cost and so on, and decide on a system architecture. A signalling system based on information technology will be introduced more actively in the future. Therefore, a systematic method of evaluating a signalling system will be more important in the development phase.

The relationship between Reliability, Availability, Maintainability and Safety are described in the RAMS standard. According to the definition, an availability and safety are the main indicators. As for safety, various technologies have been developed and applied, and signalling systems have an excellent safety record. In recent years, safety analysis is generally carried out by some systematic methods (e.g., FTA and FMEA), with appropriate measures being applied to reduce risks. On the other hand, there is no established concrete method applicable to comprehensive evaluation of the signalling system concerning availability. Furthermore, there is a growing attention to the stability of railway transport.

Therefore, we assume that any system based on proven safety technologies and safety analysis achieves a sufficient level of safety, and have focused on availability, which consists of reliability and maintainability.

3 Evaluation Method

3.1 Basic Concepts

There are several measures to express an availability. For example, it is expressed as the ratio of Mean Time Between Failure (MTBF) to sum of MTBF and Mean Time To Repair (MTTR). Although this definition expresses a performance of each equipment, the relationship between equipment failures and their impact on railway traffic service is not clear. Therefore, we propose a method to evaluate system availability based on the following concepts.

1. Estimating the impact on train operation caused by an equipment failure (a component of signalling system) .
2. Converting the impact to a loss (cost).

The proposed method will be useful to support decisions on the system architecture adapting to conditions of a line or station, making it possible to evaluate the system's life cycle cost by adding the loss to the initial and maintenance costs.

Among those three costs, initial and maintenance costs can be estimated if the system architecture is defined. However, the loss caused by failures should be derived considering the track layout in the station, train schedule and other conditions. The method to estimate the loss is described in the following section.

3.2 Method to Estimate the Loss Caused by Failures

The basic procedure is illustrated in Fig.1. We consider the signalling system in a single station. When an equipment (e.g., signal, track circuit, point machine and interlocking device) fails, there are various types of impacts. In some cases, the train cannot run at all, while in other cases the train can run through the station via an alternative route. In the latter case, although a train headway may be longer than usual, service suspension can be prevented. These type of impact can be defined for each failure case in advance.

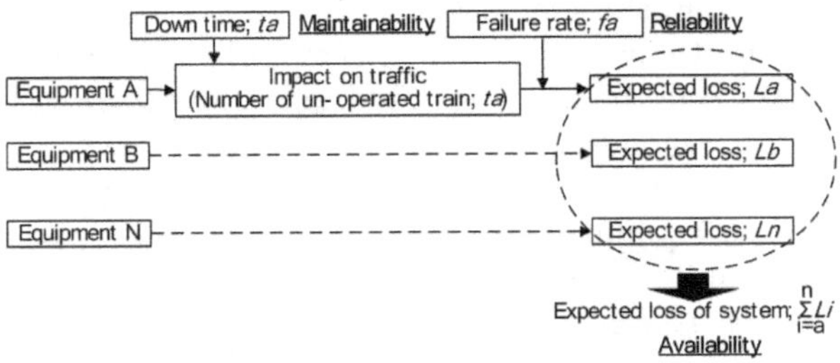

Fig. 1. A Basic Evaluation Procedure

In the first step, the number of trains which cannot be operated during a failure is estimated for each equipment based on pre-defined conditions. A duration of the failure is one of the conditions, and corresponds to an indicator of maintainability. The loss of fare revenue caused by failure is estimated from the difference between planned and actual operated number of trains. The outline of estimating the loss of each case is shown in Fig.2.

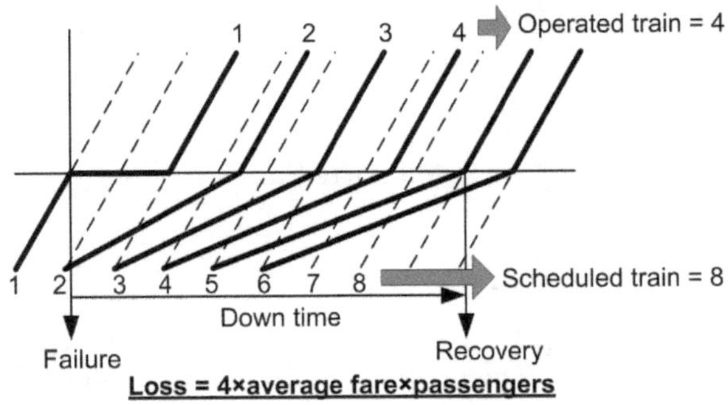

Fig. 2. Outline of Estimating the Loss

In order to obtain the loss, we need to estimate the movement of trains under various status of a train operation. The status during a failure depends on which equipment fails and the train routes related to the failed equipment. Several examples are described in section 4.1.3. Since it will be difficult to estimate the movement only by a desk study, we use the train traffic simulator to perform detailed estimation. This simulator is based on the programme which was developed for evaluating a signalling regarding a minimum headway, recovery performance from delay and so on. The original programme simulates movements of trains in reflection of train control and interlocking function. In the new simulator, a downtime and impact on a route control (e.g., all trains should stop, train runs through alternative route) can be set for each equipment. The number of trains which can be operated during the failure is obtained from the simulation.

In the next step, the expected loss is derived by a product of the loss of fare revenue obtained in the first step and the failure rate of each equipment. The failure rate corresponds to an indicator of reliability.

The expected loss of the whole system is obtained by summing the expected loss for each equipment. The resulting loss reflects factors of maintainability and reliability, and expresses the risk level of a train operation service. Therefore, it is a kind of availability indicator.

4 Case Study of Applying the Method

4.1 Preconditions

Model of Line and Station We assume three types of model stations, as indicated in Fig.3 and Table 1. Station A and Station B are common models of high density lines, and Station C is a model of suburban lines.

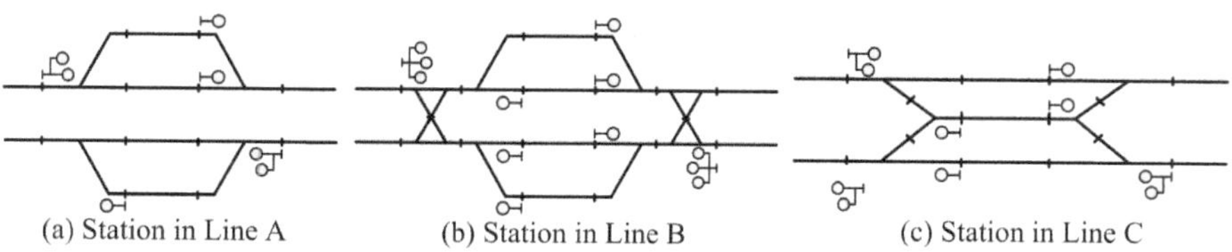

Fig. 3. Models of Stations

Model of Signalling System We have assumed three types of signalling systems for the case study. The model systems are outlined in Fig.4. The first one is a conventional system consisting of track circuits, wayside signals, etc. The second is a train control system based on radio communications, similar to CARAT [1] or ATACS [2] in Japan.

The third one is a future system based on a new concept. We performed basic experiment on the new system and have confirmed its feasibility. The system is outlined in Fig.5 and has the following features.

Table 1. Line Caracteristics of each Model Station

Model line	Train headway	Dwell time at a station
Line A Line B	120 seconds	60 seconds (A Main track and side track are used alternately.)
Line C	300 seconds	30 seconds

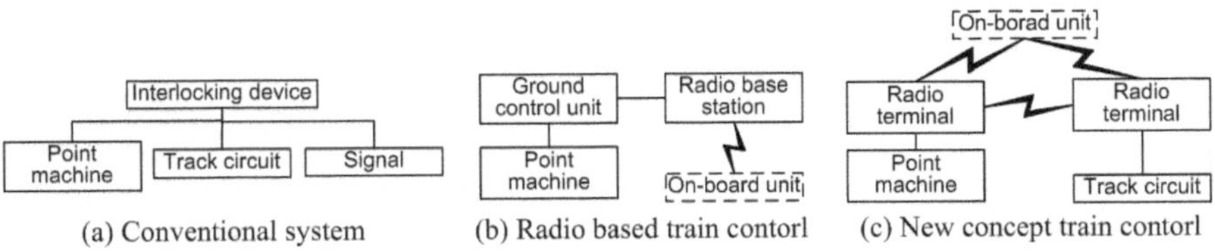

Fig. 4. Models of Systems

- Field equipment (track circuits, point machines and on-board control unit existing in a station area) have radio transmission interfaces and are connected by radio link each other.
- Interlocking function is performed based on radio communications between the equipment, and is not performed by a centralised device.
- Track circuits and point machines manage their status and requests from the train which are related to on-board control unit ID.
- The train transmits route setting requests to track circuits and point machines.
- The point machine controls itself based on the request if safety is confirmed , and transmits the answer to the train.
- The track circuit transmits movement authority to the train based on the status of train existence and point machine within its area.

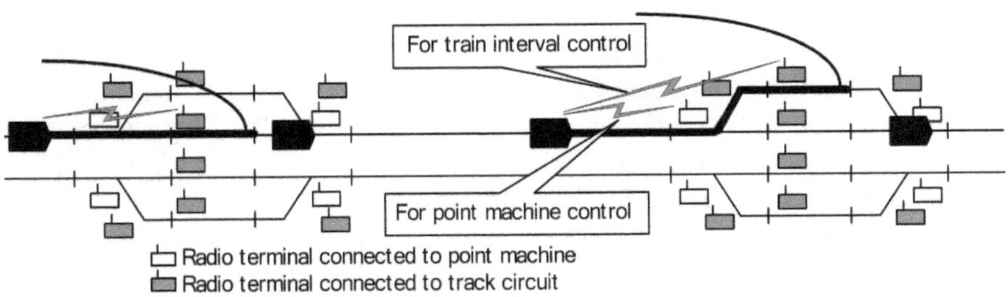

Fig. 5. Outline of new concept system

This system performs a function of conventional interlocking device by data communications between field equipment capable of autonomous action.

In a conventional system, all train operation will be stopped if an interlocking device fails. However, in this system, impact of a failure may be restricted in many cases because the interlocking function is not performed by centralised unit.

Failure of equipment We have estimated the loss at Station 13 (Line A, Line B) and Station 7 (Line C). The train operation status under failure is defined in advance. Two examples are given in Fig.6.

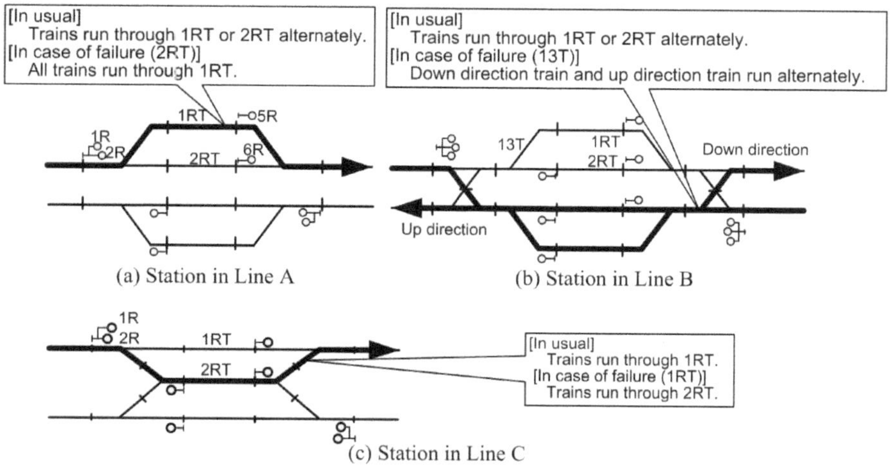

Fig. 6. Examples of Alternative Routes under Failure

The failure rates and down times of each equipment are assumed as listed in Table 2 and Table 3. The values in Table 2 are estimated from times of suspended train operation due to failures of signalling systems for about five years, which are recorded in the Railway Safety Database managed by our institute. However, the radio based system and the new system do not have actual records. For the system based on radio communications, the value for a point machine is the same as in Table 2, and the values for the ground control unit and radio base station are the same as those of an interlocking device in Table 2. In Table 3, the value for the radio terminal is assumed considering that it does not have an interlocking function of a whole of a station and its hardware may not be as complex as the ground control of radio based system.

Fig.7 shoes the results of the traffic simulation of the line B in the case where the track circuit 13T fails (Fig.6 (b)). In this case, the train headway becomes 177 seconds while the trains are operated using alternative routes. The number of trains that can run through the station depends on the down time of each model system, and is approximately from 50 to 70% of the

Table 2. Failure Rates and Down Times of Conventional System

	Interlocking device	Signal	Track circuit	Point machine
Failure rate (/h)	$1.5 \cdot 10^{-6}$	$1.0 \cdot 10^{-7}$	$3.2 \cdot 10^{-7}$	$5.9 \cdot 10^{-7}$
Down time (h)	3.28	0.78	1.39	0.60

Table 3. Failure Rates and Down Times of New Concept System

	Radio terminal	Track circuit	Point machine
Failure rate (/h)	$7.4 \cdot 10^{-7}$	$3.2 \cdot 10^{-7}$	$5.9 \cdot 10^{-7}$
Down time (h)	2.00	1.00	1.00

schedule. In the case of Fig.6 (a), the train headway becomes about 140 seconds, and the rate of operated train is about from 60 to 70%.

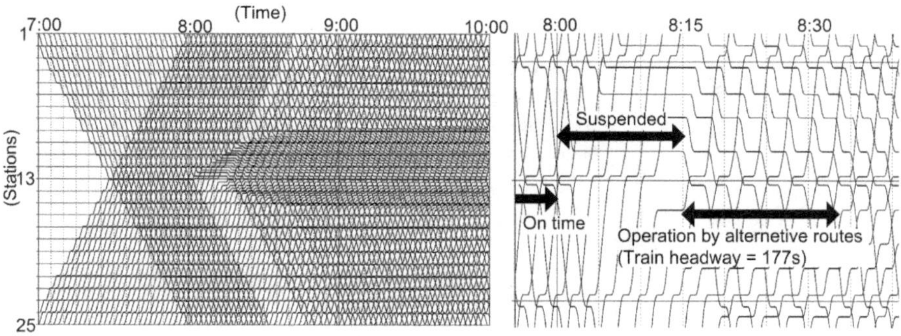

Fig. 7. Result of Simulation (In case of failure; 13T in Station 13 of Line B)

These simulations were performed for each failure case, and the expected loss for 20 years caused by equipment failure were estimated for the combination of the model lines and systems. The results are summarised in Fig.8 (a) and the sum of the loss, initial cost and maintenance cost in Fig.8 (b). They are expressed by relative values, with the assumption that the initial cost to introduce a conventional system into Line A is one.

These results indicate that the loss for the new concept system is about 1.3 times of the conventional system in Line A, and about 2.1 times in Line B. On the other hand, it is about 0.58 times in Line C. The new concept system has no losses caused by the failure of signals and the interlocking device, because these equipment do not exist. Especially, the loss caused by the interlocking device is more than 50% of the total loss in the conventional system. However, the loss caused by the failure of a newly introduced radio terminals is greater than the reduction obtained by omitting the interlocking device. The supposed reason for this is that the number of terminals is comparatively large, the assumed down time (Table 3) is longer than for other equipment, and the failure rate is a little bit higher. In contrast, the elimination of the interlocking device contributes to the reduction of the total loss in Line C because the number of equipment is comparatively small and the train headway is longer than Line A or B. Although within the assumption in this paper, in order to apply the mew system to high density lines, it is required to reduce the failure rate or down time of the radio terminal. For example, if the down time of the terminal is reduced to 50% of the value in Table 3, the ratio of total loss of the new system to the conventional system becomes about 0.80 in Line A.

These results are examples and it may change according to the conditions such as a method to estimate the loss and failure rates of equipment. However, proposed method will contribute to evaluate the signalling systems from a RAMS viewpoint and to construct appropriate system adapting to conditions of railways. Furthermore, it also enables to derive requirements to equipment from the conditions of train operation.

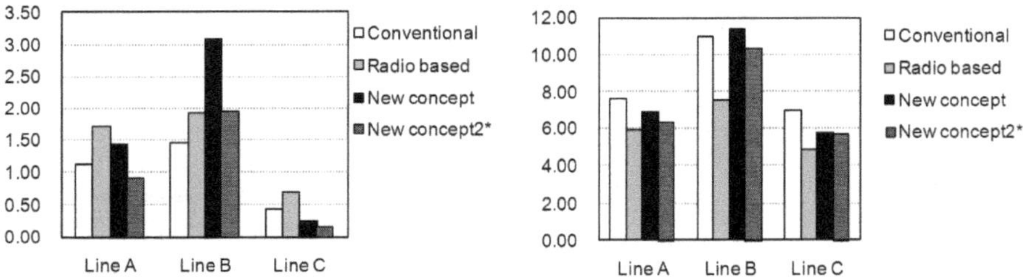

(a) Total Loss Caused by Failure (b) Sum of Loss, Initial and Maintainance costs

*New concept2 : In case that the down time of a radio terminal reduce to 50% (1 hour)

Fig. 8. Results of Estimated Loss and Costs

5 Conclusion

We proposed an evaluation method for railway signalling system from a RAMS point of view. It will be useful to support efficient designing and selection of an appropriate system which achieves the required availability, and will be able to contribute to effective system investment.

We are planning to study the RAMS performance of the signalling system according to various condition (e.g., train density, failure rate and down time).

References

1. Hasegawa, H.,: Development Steps in Application of CARAT, Quarterly Report of RTRI, Vol.36, No.1, pp.4-7 (1995)

2. Baba, Y., Hattori, T.,: Japan points the way forward, IRJ, vol.48, No.10, pp.32-33 (2008)

Model Checking Interlocking Control Tables

Alessio Ferrari[2,1], Gianluca Magnani[1],
Daniele Grasso[1], and Alessandro Fantechi[1]

[1] University of Florence (DSI)
[2] General Electric Transportation Systems (GETS) - Florence

Abstract. A challenging problem for model checking is represented by railway interlocking systems. It is a well known fact that interlocking systems, due to their inherent complexity related to the high number of variables involved, are not amenable to automatic verification, typically incurring in state space explosion problems. The literature is however quite scarce on data concerning the size of interlocking systems that have been successfully proved with model checking techniques. In this paper we attempt a systematic study of the applicability bounds for general purpose model checkers on this class of systems, by studying the typical characteristics of control tables and their size parameters. The results confirm that, although small scale interlocking systems can be addressed by model checking, interlockings that control medium or large railway yards can not, asking for specialized verification techniques.

Keywords: Railway Interlocking, Model Checking Interlocking, Control Tables Verification

1 Introduction

In the railway signaling domain, an *interlocking* is the safety-critical system that controls the movement of trains in a station and between adjacent stations. The interlocking monitors the status of the objects in the railway yard (e.g., points, switches, track circuits) and allows or denies the routing of trains in accordance with the railway safety and operational regulations that are generic for the region or country where the interlocking is located. The instantiation of these rules on a station topology is stored in the part of the system named *control table*, that is specific for the station where the system resides [12]. Control tables of modern computerized interlockings are implemented by means of iteratively executed software controls over the status of the yard objects.

Verification of correctness of control tables has always been a central issue for formal methods practicioners, and the literature counts the application of several techniques to the problem, namely the Vienna Development Method (VDM) [5], property proving [2] [4], Colored Petri Nets (CPN) [1] and model

E. Schnieder, G. Tarnai (eds.), *FORMS/FORMAT 2010*,
DOI 10.1007/978-3-642-14261-1_11,

checking [14] [10] [11]. This last technique in particular has raised the interest of many railway signaling industries, being the most lightweight from the process point of view, and being rather promising in terms of efficiency. Nevertheless, application of model cheking for the verification of safety properties has been successfully performed only on small case studies, often requiring the application of domain-related heuristics based on topology decomposition. The literature is however quite scarce on data concerning the size of interlocking systems that have been successfully proved with model checking techniques. This is partly due to confidentiality reasons, and partly to the fact that the reported experiences refer to specific case studies, with a limited possibility of scaling the obtained results to larger systems.

We have therefore decided to investigate more systematically the actual applicability bounds for widely used model checkers on this class of systems, by studying the typical characteristics of control tables and their size parameters. In particular, we have chosen to compare the performances of two popular general purpose model checkers, the symbolic model checker NuSMV [3] and the explicit model checker SPIN [8]. A test set of generic control table models of increasing complexity has been defined and translated into the input language of the two tools, and generic safety properties have been proved on them by the two model checkers. The results have confirmed that the bound on the size of the controlled yard that can be safely addressed by the two tools is still rather small, making general purpose model checking tools not usable for medium and large scale interlockings. Specifically optimized verification techniques should be studied to allow the verification of such complex systems.

2 Interlocking Systems Representation

In Relay Interlocking Systems (RIS), currently installed and operating in several sites, the logical rules of the control tables were implemented by means of physical relay connections. With Computer Interlocking Systems (CIS), in application since 30 years, the control table becomes a set of software equations that are executed by the interlocking. Since the signaling regulations of the various countries were already defined in graphical form for the RIS, and also in order to facilitate the representation of control tables by signaling engineers, the design of CIS has usually adopted traditional graphical representations such as ladder logic diagrams [4] [9] and relay diagrams [7]. These graphical schemas, usually called *principle schemata*, are instantiated on a station topology to build the control table that is then translated into a program for the interlocking.

As pointed out in [4], the graphical representations and the related control tables can be reduced to a set of boolean equations of the form $x_i := x_j \wedge \ldots \wedge x_{j+k}$, where $x_j \ldots x_{j+k}$ are boolean variables in the form x or $\neg x$. The variables represent the possible states of the signalling elements monitored by the control table: system input, output or temporary variables. The equations are conditional checks over the current and expected status of the controlled elements.

In order to give a metric to the dimension of the problem in terms of parameters of the control tables, we define the *size* of a control table as the couple (m, n), where m is maximum number of inter-dependent equations involved, that means equations that, taken in pairs, have at least one variable in common, and n is the number of inputs of the control table. We consider only inter-dependent equations because, if there are sets of equations that are independent, they can be verified separately, and slicing techniques such as the ones presented in [13] and [14] can be adopted on the model to reduce the problem size. In our experiments we basically consider control tables that have been already partitioned into slices (the size value of a control table is intuitively the one of its maximal slice).

Correctness of control tables depends also on their model of execution by the interlocking software. In building CIS, the manufacturers adopt the principle of *as safe as the relay based equipment* [1], and often the implemented model of execution is very close to the hardware behaviour. According to the semantics of the ladder diagrams traditionally used for defining the control tables, we have chosen a synchronous model with global memory space where variables are divided into input, output and latch (i.e., local)[4]. The model of execution is a state machine where equations are executed one after the other in a cyclic manner and all the variables are set at the beginning of each cycle and do not change their actual value until the next cycle. This is a reasonable generic paradigm for centralized control tables.

Timer variables are not considered in our models, since they are normally related to functional requirements of the system (e.g., the operator shall press the button for at least 3 seconds to require a route). Safety requirements such as the ones considered in this study are normally independent from timers. An intuitive argument in support of the fact that timers are not used to implement safety functions is that, in traditional RIS, timers were implemented by means of capacitors: these are components that have a rather high failure rate, making them unsuitable for safety functions.

We have developed a tool that generates a set of equations coherent with this model of execution, expressed as models suitable for automatic verification with NuSMV (see sect.4) or SPIN (see sect.5), and which represent typical control tables of parametric size.

3 Safety Requirements

Given a control table representation we want to assess that its design is correct. In the proposed experiment, we need to check that safety properties are verified, and this represents the worst case for a model checker: explicit and symbolic model checkers are challenged by verification of safety properties, since, in order to show their correctness, they have to explore the entire state space, or its symbolic representation. Safety requirements typical of signaling principles are normally expressed in the principle schemata or in the regulations. This kind of properties have shown to be representable in Computation Tree Logic (CTL) in the CTL-AGAX form: $AG(p \rightarrow AXq)$, where p and q are predicates on the variables of our model [6]. CLT-AGAX formulae have an equivalent Linear Temporal Logic (LTL) representation. The formula $AG(p \rightarrow AXq)$ can be expressed in LTL syntax as a LTL-GX formula of the form $G(p \rightarrow Xq)$. Intuitively, they represent fail-safe conditions, i.e., events that should happen on the next state if some unsafe condition occurs. One of the typical safety properties that is normally required to be verified is the *no-derailing property*: while a train is crossing a point the point shall not change its position. This typical system level requirement can be easily represented in the AGAX form [14]:

$$AG(occupied(tc_i) \wedge setting(p_i) = val \rightarrow AX(setting(p_i) = val))$$

whenever the track circuit tc_i associated to a point p_i is occupied, and the point has the proper setting val, this setting shall remain the same on the next state.
In order to force the worst-case full state space exploration, our test set has been designed on purpose to satisfy given properties expressed in CTL-AGAX (or LTL-GX) form, and model checking has been performed using these properties as formulae. Though not clearly evident, also for symbolic model checking we have experienced that satisfied invariants are the hardest problem.

4 NuSMV

NuSMV is an open-source symbolic model checker that provides the user with both Binary Decision Diagrams (BDD) based implicit model-checking and SAT solver based bounded model checking. Properties are encoded in CTL in the first case, and in LTL in the second case. Since we focus on the verification of safety properties, we need to be sure that every single reachable state is analyzed by the model checker; for this reason, we have not used NuSMV bounded model checking [1].

In NuSMV, the state is represented by the value of state variables. The next state is computed by first calculating the next values of state variables and then, atomically, updating all the state variables. This behaviour of the model checker is compliant with the chosen model of execution. Every equation is hence evaluated in sequence but the outputs are updated at the end of the whole evaluation phase. This behaviour permits to be free from the order of evaluation of the equations.

NuSMV supports open models, that means that it computes all possible input variable values automatically: in its internal modeling language the

[1] Altough there are techniques that are able to guarantee in some cases the full exploration of the state space with bounded model checking, these have not been used in these initial experiments and will be the subject of further experiments.

keyword IVAR must be used for such variables. Input variables do not contribute to expand the state space of the system, but influence the number of reachable states. The variables under the keyword VAR are indeed state variables: the value of each of them in the model is determined by the evaluation of the conditions. In Fig. 1(b) is represented an extract of a NuSMV model used for our case study.

An example of a CTL-AGAX property that is verified on this SMV model is: AG(out0=0 → AX(out1=0)).

5 SPIN

SPIN is a generic verification system based on explicit model checking. It performs a full search in the state-space to find whether or not a given set of system properties, that are expressed using Linear Temporal Logic (LTL), are satisfied. The systems are modeled by the verification language PROMELA (PROcess MEta LAnguage), a C-like language, that provides instruments especially for the modelling of distributed asynchronous systems.

In this section we will distinguish between input and state variables of the system that is modeled using PROMELA: this distinction is just conceptual, since every variable in the model is in fact a state variable. SPIN updates the state variables on the fly, after the evaluation of the conditions of each equation; since we want to model a state machine compliant with the chosen execution model, we had to add to the PROMELA model a number of temporary variables equal to the number of actual state variables, that are updated only after the evaluations of all the equations, at the end of each processing step.

A LTL-GX property, corresponding to the CTL-AGAX one given for NuSMV, that is verified on this PROMELA model is:
□(out0=0 → X(out1=0)).

An example of model fragment written in PROMELA, corresponding to the one shown for NuSMV, is represented in Fig. 1(b). Since SPIN can only verify closed models, we had to model the environment behaviour in the PROMELA model: in order to model the non-determinism of the input variables values, we need to insert at the end of the PROMELA model an *if statement* for each input variable, as represented in the Fig. 1(b).

6 Results and Discussion

In order to investigate the actual applicability bounds for the model checking of interlocking systems, the experiments were performed on generated models with different equations-inputs ratio. For each combination of inputs and equations at least three different generic control table models were generated and tested, in order to be able to avoid erroneous positive results caused by the generation of trivial equations: if at least one of the three model is

```
MODULE main
IVAR
in0:boolean;
in1:boolean;
[..]
VAR
out0:boolean;
out1:boolean;
ASSIGN
[..]
next(out0):= !in2 & in3 |
             !in4 & in5:1;
             1:0
next(out1):= !in2 & in3 |
             !in6 & out0:1;
             1:0;

[..]
```

(a)

```
[..]
if
::((in2==0 & in3==1) |
   (in4==0 & in5==1))->tmp0=1;
::else tmp0=0;
fi;
if
::((in2==0 & in3==1) |
   (in6==0 & out0==1))->tmp1=1;
::else tmp1=0;
fi;
[..]
out0=tmp0;
out1=tmp1;
[..]
if
:: true -> in1=0;
:: true -> in1=1;
fi
[..]
```

(b)

Fig. 1. NuSMV and SPIN model extracts

not verifiable the whole class of models with the same ratio is considered not-verifiable.

Fig. 2 shows the results of the verification runs using NuSMV. The *non-verifiable* cases was related either to memory exhaustion or to several hours of execution without any answer (a threshold of 36 hours has been chosen)[2].

The upper bound of applicability in the case of 1\10 ratio and 1\5 ratio is almost the same, approximately 70 equations. Otherwise, considering a ratio of 1\2, the number of different inputs causes the increase of the degrees of freedom and the consequent explosion of the reachable states: the computational time is considerably increased and the upper bound of applicability decreases to 60-65 equations. In the examined cases, using different optimization settings for NuSMV has not produced significant performance improvements. Fig. 3 shows the experimental results obtained by the execution of SPIN on the same dataset used in the experiments with NuSMV. We observe that increasing the ratio between inputs and expressions causes SPIN not to conclude the verification in a fair time or, for higher values, to crash due to the massive usage of system memory. This behaviour can be tracked to the fact that input variables are actually state variables: so increasing inputs causes a state-space explosion. A similar analysis can be performed for the equations, since every new equation brings a new state variable for the system. It was found that the upper limit for the applicability of SPIN to an interlocking problem is about 80 equations and 20 inputs without using any memory

[2] The verification were run on a pc with 4.0 GB of ram and a 2.4 GHz core.

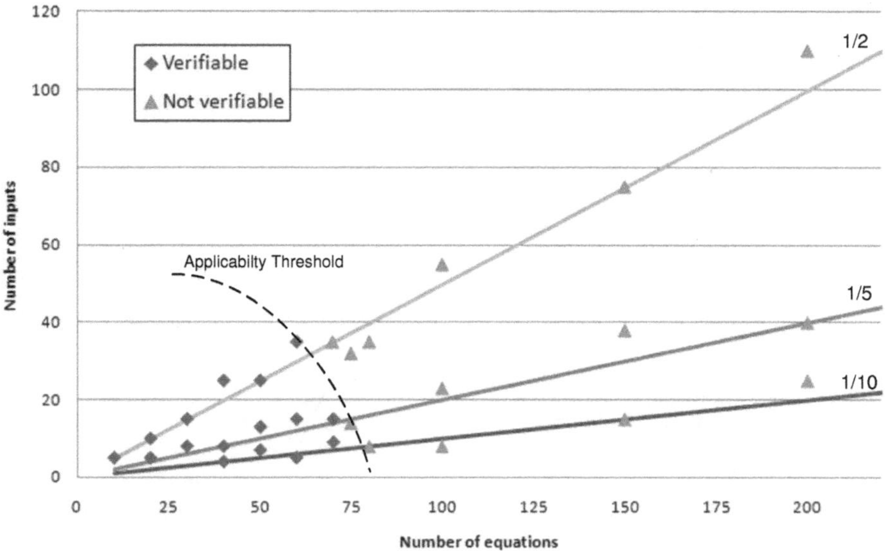

Fig. 2. NuSMV results

oriented optimization. SPIN offers several optimization strategies (e.g., hashcompact, bitstate hashing), and, according to our experiments, the one that resulted in major benefits for our case study is the one called Minimized Automata, that consists in the construction of a minimal deterministic finite state automaton. This optimization allows a significant memory usage reduction, increasing, on the other side, the time needed for the execution. The usage of such optimization increases the limit of applicability to about 100 equations and 60 inputs.

The results obtained with our approach show that the model checking applied to an entire interlocking system of medium size (normally some hundreds of equations) is already unfeasible. We have however to note that the results are given on sets of strongly inter-dependent equations: an interlocking system where slicing techniques can be applied to separate sets of interdependent equations can be much larger. Clearly, slicing can be applied only if the actual topology of the tracks layout and the interlocking functionality do separate concerns about different areas of the layout, with little interactions among them. Considering the real world interlocking of [13], we can attempt to verify the correctness of only the smallest slice identified in the paper (4 signals, 7 track circuits and one switch), while the bigger slices might outrun the capability of the considered model checkers. Nevertheless, the entire interlocking is a large size one, and normally medium size interlocking present smaller slices, making the problem of their correctness addressable by model checking.

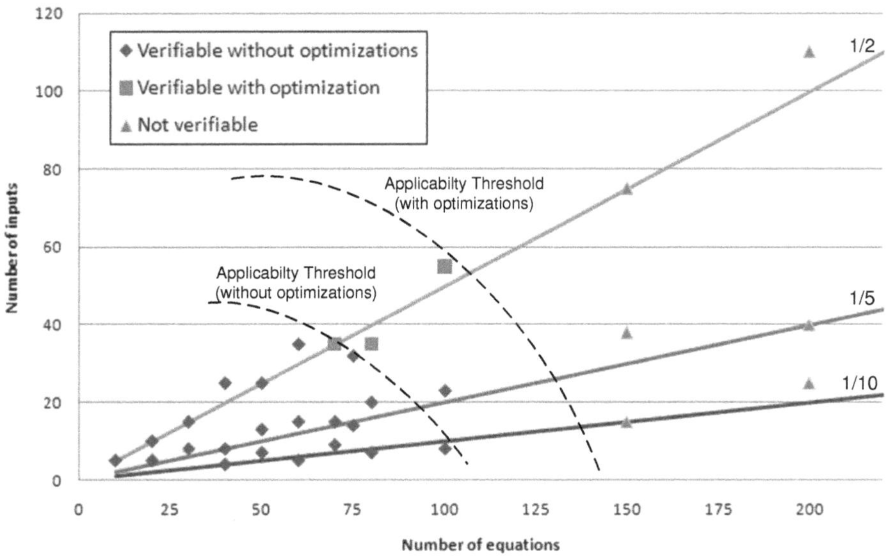

Fig. 3. SPIN results

7 Conclusion

We have studied the application of general purpose model checkers to railway interlocking systems, with the aim to define the upper bounds on the size of the problem that can be effectively handled. For this purpose, we have defined the size parameters of an interlocking systems on the basis of its control tables, and we have conducted experiments on purposely built test models of control tables with the NuSMV and SPIN model checkers. The results have confirmed that, although small scale interlocking systems can be addressed by model checking, interlockings that control medium or large railway yards can not with general purpose verification tools. In order to increase size of tractable interlocking systems several directions will be pursued in future work, such as:

- automated application of slicing;
- safe assumptions on the environment, that can tailor the input space to the one actually encountered in practice;
- considering the use of specialized model checkers for PLCs[15];
- use of proper variants of SAT-based bounded model checking that are able to efficiently prove safety properties.

References

1. Anunchai, S.V.: Verification of Railway Interlocking Tables using Coloured Pertri Nets. Proceedings of the 10th Workshop and Tutorial on Practical Use of Coloured Petri Nets and the CPN Tools (2009)

2. Boralv, A.: Formal Verification of a Computerized Railway Interlocking. Formal Aspects of Computing **10** (1998) 338–360
3. Cimatti, A., et al.: NuSMV 2: An OpenSource Tool for Symbolic Model Checking. CAV 2002, LNCS 2404, 359–364
4. Fokkink , W., Hollingshead, P.: Verification of Interlockings: from Control Tables to Ladder Logic Diagrams. 3rd FMICS Workshop (1998) 171–185.
5. Hansen, K.M.: Formalizing Railway Interlocking Systems. Proceedings of the 2nd FMERail Workshop (1998)
6. Haxthausen, A.E., Peleska, J.: Formal Development and Verification of a Distributed Railway Control System. Proceedings of FM'99, LNCS 1709 (1999) 1546 – 1563
7. Haxthausen, A.E.: Developing a Domain Model for Relay Circuits. International Journal of Software and Informatics (2009) 241–272
8. Holzmann, G.J.: The SPIN Model Checker : Primer and Reference Manual. Addison-Wesley Professional (2003)
9. Kanso, K., et al.: Automated Verification of Signalling Principles in Railway Interlocking Systems. ENTCS **250** (2009) 19–31
10. Mirabadi, A., Yazdi, M.B.: Automatic Generation and Verification of Railway Interlocking Control tables using FSM and NuSMV. Transport Problems : an International Scientific Journal **4** (2009) 103–110
11. Pavlovic, O., Ehrich, H.: Model Checking PLC Software Written in Function Block Diagram. 3rd ICST (2010) 439–448
12. Tombs, D., et al.: Signalling Control Table Generation and Verification. Proceedings of the Conference on Railway Engineering (2002)
13. Winter, K., Robinson, N.J.: Modeling Large Railway Interlockings and Model Checking Small Ones. Proceedings of the 26th Australasian Computer Science Conference **35** (2003) 309–316
14. Winter, K., et al.: Tool Support for Checking Railway Interlocking Designs. Proceedings of the 10th Australian Workshop on Safety Critical Systems and Software (2006) 101–107
15. Schlich, B., Brauer, J., Wernerus, J., Kowalewski, S.: Direct Model Checking of PLC Programs in IL. Proceedings of DCDS (2009) to appear

Reliability of the IP Network-based Signal Control System and the Integrated Logical Controller

Takashi Kunifuji[1], Yoshinori Saiki[1], Satoru Masutani[2], and Masayuki Matsumoto[2]

[1] Research Development Center of JR East Group, East Japan Railway Company, 2-479 Nissin-cho, Kita-ku, Saitama, 331-8513 Japan
{y-saiki, kunifuji}@jreast.co.jp

[2] Electrical & Signal Network Department, East Japan Railway Company, 2-2-2, Yoyogi, Shibuya-ku, Tokyo, Japan
{s-masutani, m-matsumoto}@jreast.co.jp

Abstract. East Japan Railway Company developed the IP Network-based Signal Control System, which was introduced into service in February 2007, with an innovative change of the control techniques for wayside signaling devices from power control through conventional metallic cables to digital information control through an optical network. This system is intended to improve the quality of construction of signaling devices by reducing the wiring work that is required for a large number of metallic cables and simplifying the operation tests. In this system, reliability is improved by using the duplex transmission lines, including the optical network. In this paper, we describe reliability of the IP Network-based Signal Control System and discuss application to the Integrated Logical Controller that is currently under development.

Keywords: PON, Internet Protocol, Railway Signal, Equipment Integration

1 Introduction

In a conventional railway signaling system, railway signals and motors are controlled by interlockings through metallic cables. Because there are many metallic cables, the wiring work and the operation test become problems. Therefore, East Japan Railway Company has developed the IP Network-based Signal Control System to innovate a change in the control techniques for wayside signaling devices from power control through conventional metallic cables to digital information control through an optical network, and introduced it into service in February 2007. The goal of this system is to improve the quality of construction of signaling devices by reducing the wiring work that is required for a large number of metallic cables, and by simplifying the

E. Schnieder, G. Tarnai (eds.), *FORMS/FORMAT 2010*,
DOI 10.1007/978-3-642-14261-1_12, © Springer-Verlag Berlin Heidelberg 2011

operation tests in the field. This system improves reliability by using duplex transmission lines including the optical network. In this paper, we describe reliability of the IP Network-based Signal Control System and discuss its application to the Integrated Logical Controller, which is under development.

2 Overview of the IP Network-based Signal Control System

Figure 1 shows the system configuration of the IP Network-based Signal Control System. [1],[2] The roles of each device are also shown below.

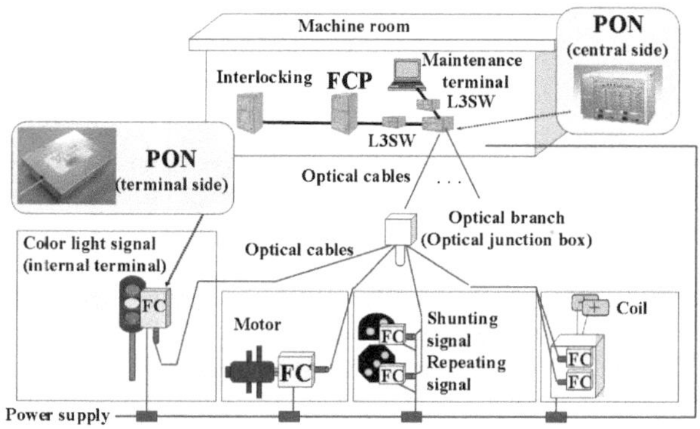

Fig. 1. System configuration of the IP Network-based Signal Control System

2.1 Field Object Controlled Processor (FCP)

The FCP decides what control information is to be sent to the target FCs (see 2.3) based on the control information from the Interlocking and indication information on the related signal device, and transmits the control message to all FCs by broadcast transmission. The FC extracts the control command that corresponds to it from the control message, and controls the signal devices. The control result is sent back to FCP by unicast transmission. The process concerning the safety of the control information transmission is in accordance with IEC62280-1. [3], [4].

2.2 Network

The network uses Ethernet (100BASE-TX) in the machine room, and PON (Passive Optical Network) in the field. PON fits for constructing the network for the signal control devices at the wayside in the following ways.

- Because the optical fiber is capable of two way communication by one fiber, the number of lines can be reduced and the construction becomes simple.
- Because the optical fiber is capable of branching by an optical coupler near the signal devices, the amount of construction of the optical fiber can be reduced. Moreover, no power supply is necessary for the coupler.
- Fixed bandwidth of a transmission signal can be allocated in each node and it fits for real-time communication.

2.3 Field Controller (FC)

The FC is a failsafe I/O terminal built into the body of a signal device or the wayside box along a railway track. However, the FC does not have any interlocking logic.

3 Reliability of the IP Network-based Signal Control System

Figure 2 shows the network configuration of the IP Network-based Signal Control System. This system needs to maintain reliability as the electronic devices are installed along the wayside in a difficult environment, and various measures have been taken such as multiple systems or multiple data paths. Concretely, the hardware of FCP and FC, and the transmission line have doubled compositions. Moreover, it is difficult to ensure the reliability indicated in the requirement specification with a single composition, because general-purpose units are used for the transmission devices. Therefore, not only the transmission line is doubled but also each composition of the FCP sends the control message to both compositions of the FC and each composition of the FC sends back the control result to both compositions of the FCP. As a result, each composition of the FCP and the FC receives the same message twice (four times in total in both compositions). Each composition selects the correct and first received data. Wrong data detected with CRC are deleted. The reliability is improved by this procedure. The transmission between FCP and FC is transmission between fail safe computers, though the transmission line is not fail safe, and adequate reliability is ensured. (IEC62280-1) [3], [4]

In addition, the logic unit of an existing electronic interlocking uses the majority decision logic of 2 out of 3 to ensure safety, but we experienced some trouble with the device that processed the majority decision. Therefore, the double data comparison system is used this time, and a dual duplex method is adopted by using two of this doubled system for reliability. Because the switch composition is not needed when all the compositions are activated, observation of breakdown composition and method of switch composition are simple and the reliability of software is improved.

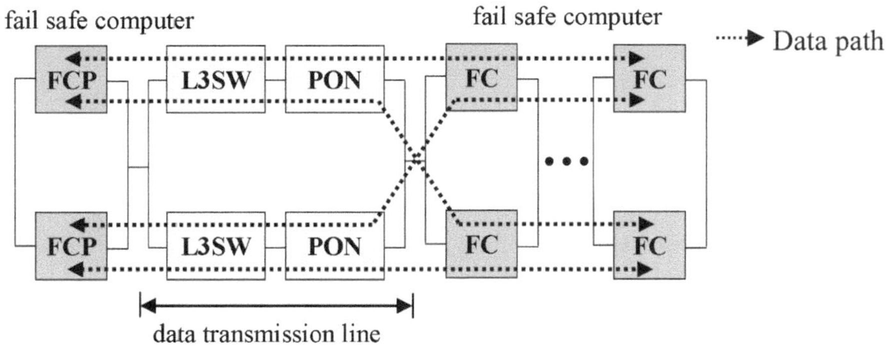

Fig. 2. Network configuration of the IP Network-based Signal Control System

4 Problem of the IP Network-based Signal Control System

The IP Network-based Signal Control System has achieved simple construction and high reliability by connecting between the logical part in a machine room (such as an existing electronic interlocking) and the wayside devices with a double optical network. However, existing electronic interlockings have an internal optical network called S-LAN. The system configuration of connecting to another optical network does not offer benefits in reliability and cost-effectiveness. Moreover, because the FCP does not have the I/O function like an existing field device controller and is specialized in logical processing, the process of FCP should be consolidated into the process of an existing logic unit of one computer (see Figure 3). Therefore, the development plan for a new Integrated Logical Controller was made.

5 Overview of the Integrated Logical Controller

The current logical part of the machine room has a complex configuration that consists of multiple logical units which cooperate for controlling the signal devices in the station yard. This complexity remains a problem in the signal control system after the development of the IP Network-based Signal Control System, because a lot of work is still required for the design, construction, and inspection. To solve this problem, we are developing the Integrated Logical Controller (LC), which covers the function of controlling all signaling devices in the station yard with one logical controller, by simplifying the system configuration and by integrating the control logic. [5],[6]

In the LC, the logical units and the control parts are integrated into one logical controller, and the connections with wayside devices are combined for transmission of information by optical cables. As a result, this feature brings easiness in design, construction and inspection because of the simplicity of

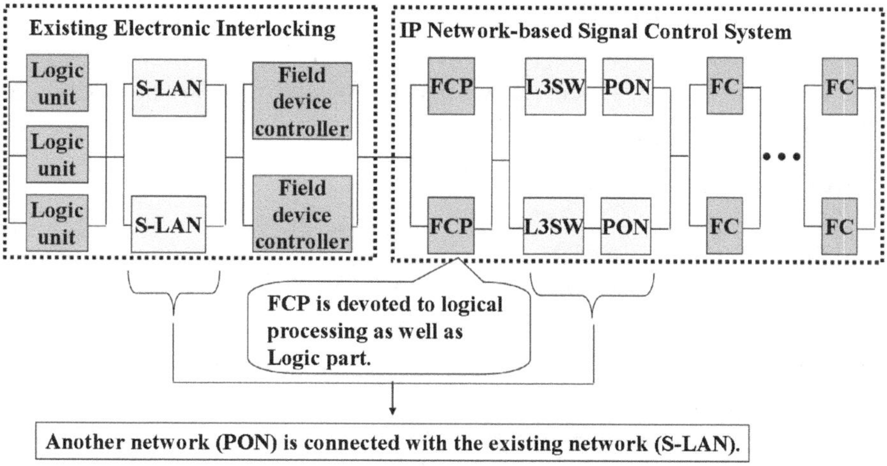

Fig. 3. Problem of the IP Network-based Signal Control System

the system configuration. Moreover, the LC achieves the improvement of reliability as an entire system by adopting a double system composition of failsafe hardware.(see Figure 4)

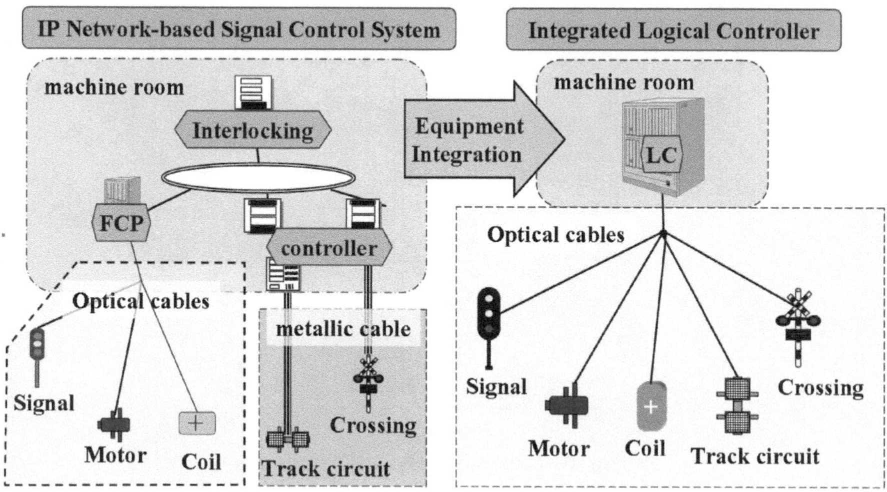

Fig. 4. Equipment Integration

6 Reliability of the Integrated Logical Controller

As a technique, it is generally known that the reliability of equipment can be improved by the following methods.

- Reduce the number of parts
- Improve the quality of parts
- Reduce mechanical parts
- Make parts mutually independent so that failure of one does not cause others to fail

In developing the LC, we focused on reducing the number of parts to improve reliability.

Figure 5 shows the reliability block chart of the IP Network-based Signal Control System and LC. The IP Network-based Signal Control System is connected in series with the existing electronic interlocking device in order to reduce the new development as much as possible. On the other hand, the FCP and logic unit are integrated to the LC, and the LC processes what the FCP and upper logic unit process. Use of the LC decreases the failure rate to about 1/5 compared with the combination of an existing electronic interlocking device and The IP Network-based Signal Control System. This is the effect of the reduction of the number of parts as the S-LAN and Field device controller are removed.

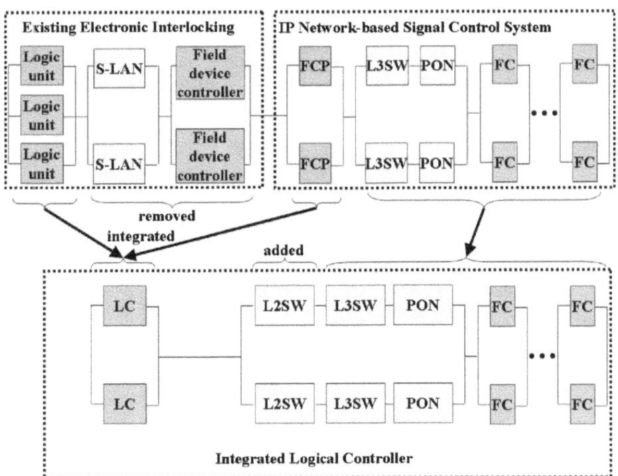

Fig. 5. Reliability Block Diagram

Moreover, the L2SW is added as one of the change points from the IP Network-based Signal Control System to the LC. The addition of the L2SW was examined as shown in Table 1.

Table 1. Examination of L2SW addition

		A	B
		without L2SW	with L2SW
1	The number of Ethernet interfaces necessary for LC	1	2
		○	△
2	The number of L2SW	0	2
		○	△
3	Transmission route between LC and FC when one L3SW breakdown	only one route	two route
		△	○
4	The addition of spare LC	difficult	easy
		△	○
Evaluation		△	○

- For the number of Ethernet interfaces, hardware and software in the LC are almost the same. In the FCP, two Ethernets cannot be achieved.
- For the number of L2SW, the disadvantage of increase in the number of devices is small because L2SW is inexpensive and has a long MTBF.
- The transmission route between LC and FC when one L3SW breaks down becomes only one route when L2SW is not added, as shown in Figure 6. On the other hand, it becomes two routes when L2SW is added, and the redundant configuration of transmission is ensured.
- For the addition of spare LC, when L2SW is not added, the re-composition of the network is needed and this is a large alteration.

From the evaluation of the above four points, we concluded that the addition of L2SW was effective.

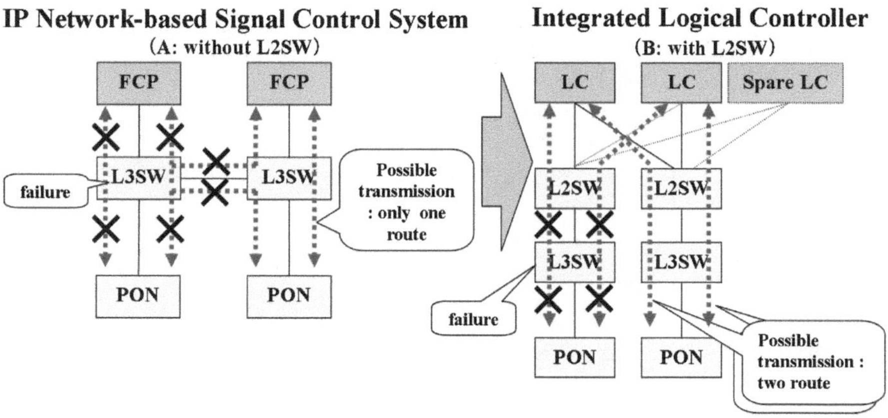

Fig. 6. Transmission route between LC and FC when one L3SW breaks down

7 Conclusion

In this paper, the reliability and the problems of the IP Network-based Signal Control System were clarified, and the application to the LC, which is currently under development, was shown to be the solution to them.

In the LC, the reduction of the parts of S-LAN and field device controllers was discussed and the transmission route between LC and FC when one L3SW breaks down was examined. As a result, we showed that LC obtains higher reliability than the IP Network-based Signal Control System.

We will continuously work on the problem extraction and find the solution to it towards the practical use of LC.

References

1. Y. Hirano, Takashi. Kato, T. Kunifuji, T. Hattori, Tamotsu Kato: Development of Railway Signalling System Based on Network Technology, IEEE SMC, (2005)
2. Y. Fukuta, G. Kogure, T. Kunifuji, H. Sugahara, R. Ishima, M. Matsumoto: Novel Railway Signal Control System Based on the Internet Technology and Its Distributed Control Architecture, IEEE ISADS, (2007)
3. IEC62280-1, International Electrotechnical Commission, 2002.
4. F.Kitahara, "ATOS system for Realization of NewTransport Operation Control," rail international, UIC, No.6, pp.14-22, 1996.
5. T. Kunifuji, T. Miura, G. Kogure, H. Sugahara, M. Matsumoto: A Novel Railway Signal Control System Based on the Internet Technology and an Assurance Technology, IEEE ADSN, (2008)
6. T. Kunifuji, K. Mori, J. Nishiyama, G. Kogure: Innovation of the Railway Signalling System Utilizing the Network and the Autonomous Decentralized Technology, IEEE ISADS (2009)

Methodology for Assessing Safety Systems Application for a Railway Hot Box Protection System*

Joffrey Clarhaut, Etienne Lemaire, and El Miloudi El Koursi

Université Lille - Nord de France,
Institut National de Recherche sur les Transports et leur Sécurité, INRETS,
BP317, 20 rue Elisée Reclus,
59666 Villeneuve d'Ascq Cedex, France
{joffrey.clarhaut, etienne.lemaire,
el-miloudi.el-koursi}@inrets.fr
http://www.inrets.fr

Abstract. This paper presents a methodology for assessing railway safety systems. This methodology is based on a SADT model of the system and on a probabilistic evaluation of each function and sub-function of this model. As an illustration, this methodology is applied to assess a specific railway protection system: the hot axle box protection system. Risks involved with a hot axle box are also presented. Finally, results obtained in the case of a high-speed rail line are analysed.

Keywords: SADT Modelling, Probabilistic evaluation, Hot Box Protection System, Railway

1 Introduction

In railway systems, safety and reliability levels of a number of existing systems are based on knowledge and existing rules of a single railway company. These rules and this knowledge may not be modelled or demonstrated. Moreover, these proprietary rules can not be used in order to define new specifications for equipment admittance. Several problems arise: how to establish safety and reliability levels when the system can not be treated using the concepts of the EN50126 standard [1]? How to check these levels and to prove that they meet the regulatory requirements of the infrastructure manager?

In this paper, we define a general method for the determination and the demonstration of safety levels. This method consists of two steps: a first step of system modeling and a second step of evaluation of the various sub-systems of the considered system.

* The present research work has been supported by the French Railway Safety Authority (EPSF). The authors gratefully acknowledge the support of this institution and reviewer's comments.

E. Schnieder, G. Tarnai (eds.), *FORMS/FORMAT 2010*,
DOI 10.1007/978-3-642-14261-1_13, © Springer-Verlag Berlin Heidelberg 2011

The protection system against hot axles boxes (also called as *hot box*) is one of these systems that are based on knowledge and existing rules. We propose to illustrate our method by evaluating this particular protection system.

The first part of this paper presents the system for detecting hot boxes and risks associated with the lack of detection. The second part of this paper presents the model proposed. Finally, a probabilistic evaluation and results obtained for high-speed lines are shown.

2 Presentation of train hot boxes

2.1 Definition and frequency

In railway systems, a *hot box* is an axle box that has reached a overheating temperature. According to [2], the axle temperature can rise to 900°C or 1000°C and can break the axle. Fortunately a broken axle is quite rare on the European network. The ERA safety performance report [3] shows, in Europe, 100 broken axles for 4225 million tr.km in 2007. Given the disparity in reporting, these results should be taken with caution. In fact, they are not completely reliable but they give an indication of the number of broken axle per million tr.km: between $1,15.10^{-2}$ and $2,36.10^{-2}$ broken axle per million tr.km.

Scientific literature also provides further informations on the number of broken axle in the United Kingdom: 1-2 broken axles per year on the U.K. network as reported by [4], an average of 1,6 broken axle per year over the past 25 years (between 1975 and 2000) on a population of 180000 axles as reported by [5]. If these numbers are reported in million tr.km in the U.K., it can be estimated, with the assumption of 20 millions kms travelled, that there are about 8.10^{-2} broken axle per million tr.km.

2.2 General risks and consequences

An overheated axle box can lead to several consequences for the train:

- it can cause a fire (but it seems unusual),
- it can lead to a rupture of the axle which give rise to two major possibilities of serious injury:
    - a derailment of the train (which is the most frequent case),
    - a loss of braking capability for the train.

2.3 Train protection systems against hot boxes

As previously shown, consequences of an overheating axle has significant consequences. The STEM (supervisory of running trains) agents [6] intend to monitor trains in order to detect overheated axles as soon as possible.

Roles and responsibilities of these agents are generally shared between railway managers and infrastructure managers. We distinguish two types of protection systems: *technical systems* and *non-technical systems.*

Non-technical systems It means all human procedures available to allow traffic safety throughout the rail network. Three types of people are distinguished:

- *the train driver* for the safety of its own train. In addition, as he meets other trains, he can inform traffic officers (dispatchers, controllers ,...) of damages he might see on these trains.
- *on board train staff* and *on board train passengers* can see anomalies such as: smokes, oil squirts, abnormal vibrations of the train, ...
- *on track staff* can also see anomalies such as: smokes, oil squirts, burning smell, ...

Technical systems It means all technical equipments that can detect a hot box. Technical equipments are called as *Hot Box Detectors* (HBD). There are two types of HBD: on board and on track detectors.

- *On board HBD* only concerns new trains. They directly inform the train driver when a hot box has been detected. These systems will be managed in Europe by the EN15437-2 standard *Railway applications - Monitoring the state of the axle boxes* which is being drafted.
- *On track HBD.* In Europe, HBD are placed along the railway. They respect the EN15437-1 standard [7]: two thermal infrared detectors at either side of the railway measure the thermal radiation of axle. Temperature readings are compared to three types of alarms as defined in [7].

3 Hot box protection system modeling

Based on the description of protection systems previously made, a SADT model of the protection system in two independent sub-systems is proposed [8]:

- *To monitor trains against hot box danger*: this function allows the detection of a hot box,
- *To manage a hot box alarm*: this function ensures the safety of train traffic after a hot box alarm.

This function is divided into four independent sub-functions: *To monitor axle boxes by train staff*, *To monitor axle boxes by STEM agents*, *To monitor axle boxes by on track HBD* and *To monitor axle boxes by on board HBD.* Due to page limitations, only the function *To monitor trains against hot box danger* and the first three sub-functions will be presented. The complete SADT model is represented by figure 1. For each sub-function, a train fitted for transport is applied in input and there is two output data: an alarm on the crossed train and then the train became unfit for transport. The control data corresponds to a visible axle box anomaly.

3.1 To monitor axle boxes by train staff

For this sub-function, three activities are distinguished:

- Train drivers of cruisers trains can perceive visually anomalies on other trains,
- Train drivers can detect problems on their own train,
- On board staff and train passengers can see anomalies on the rolling stock.

To perceive anomalies on crossed trains
This activity is represented by figure 1 (a). It needs a train driver for a cruiser train that can detect a visible axle box anomaly on other trains. This activity is represented by three parameters: the train driver attention for his driving environment (TDA), the percentage of trains supervised (Visual Inspection Quotient VIQ) and the percentage of time that the driver spends to monitor the train he meets (Train Crossed Quotient TCQ).

To detect problems on their own train
The SADT model is represented by figure 1 (b). It also needs a train driver that can detect a visible axle box anomaly on their own train. This activity is represented by three parameters: the train driver attention (TDA), the percentage of supervised wagons (Wagons Supervised Quotient WSQ) and the percentage of time that the driver spends to monitor its own train (Train Crossed Quotient TCQ).

To detect problems by on board staff and passengers
This activity is only available for passenger trains that are not empty. The SADT model is represented by figure 1 (c). It needs both passengers and crew that can detect a visible axle box anomaly. This activity is represented by four parameters: the passenger attention (PA) for their environment, the number of passengers by wagon (NP), the percentage of passengers implicated (Passengers Implication Quotient PIQ) and the crew procedural action (CPA).

3.2 To monitor axle boxes by STEM agents

The SADT model is represented by figure 1 (d). It needs STEM agents with a role in traffic safety. Several types of agents can be used: switchers, regulators, staff at stations, staff at docks, ... This activity is represented by three parameters: the STEM agents attention ($STEMA$), the STEM agent quotient (SAQ) and the number of agents available (NA).

3.3 To monitor axle boxes by on track HBD

A single on track HBD is considered in this paper. Its objective is to measure temperature of axle boxes, to compare it to alarm thresholds and, in case of an alarm, to inform the control station. The SADT model is shown by figure 1 (e). This activity is represented by five parameters: the number of detectors (NB_{HBD}), the number of failures tolerated per year (NFT), a risk factor

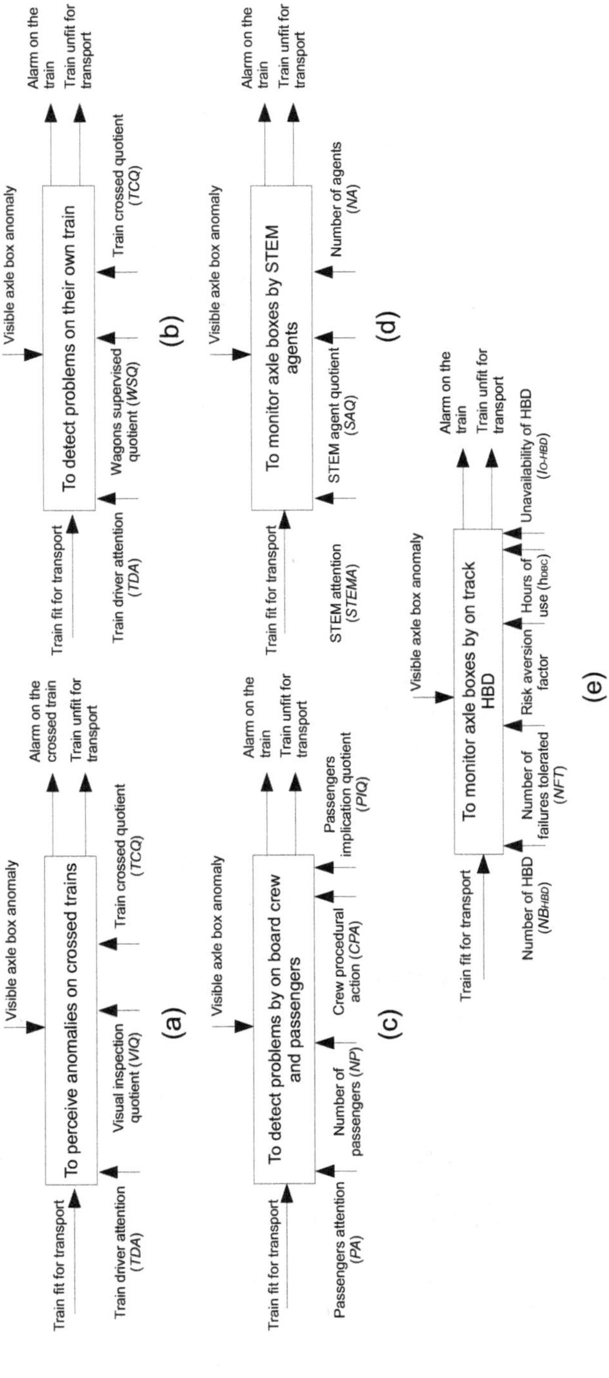

Fig. 1. SADT model of the hot box protection system

that indicates if this particular risk is estimated as *very important* or not [9], the unavailability of HBD (I_{0-HBD}) and the number of hours of operation (h_{DBC}).

4 Evaluation of the model and implementation on a high-speed line

In this section, each previous sub-function is evaluated thanks to a probabilistic approach. Then, this model is applied for a high speed line case.

4.1 Evaluation of sub-functions and activities

To perceive anomalies on crossed trains The estimated rate of non-detection of a hot box (RDH) by the train driver of a cruiser train (TCD) is represented by equation 1.

$$RDH_{TCD} = \frac{RDH_{TDA}}{VIQ \times TCQ} \tag{1}$$

- Assuming an average train driver attention for his driving environment: $RDH_{TDA} = 5.10^{-2}$ / h.
- Assuming that only 50% of cases detected by the driver of a train cruiser are visual and 90 % of the information perceived visually in the process of human perception is taken into account [10]. $VIQ = 0,9 \times 0,5 = 0{,}45$.
- Assuming that the driver spends 10% of his time to monitor the train he meets: $TCQ = 0{,}1$ (10 %).

The result of RDH_{TCD} gives a number greater than 1. It is proposed to neglect the effect of this activity in the model.

To detect problems on their own train The estimated rate of non-detection of a hot axle box by the train driver is represented by equation 2.

$$RDH_{Driver} = \frac{RDH_{TDA}}{WSQ \times TCQ} \tag{2}$$

- Assuming an average train driver attention to his driving environment: $RDH_{TDA} = 5.10^{-2}$ / h.
- The only supervised wagons of the trains are the locomotive or 20% of train: $WSQ = 1$ or 0,2.
- Assuming that the driver spends 25% of his time to monitor the state of its own train: $TCQ = 0{,}25$.

Results for this activity are: for the locomotive: $RDH_{Driver} = 2.10^{-1}$ / h and for the rest of the train: $RDH_{Driver} = 1$ / h. It is again proposed to neglect the effect of this activity in the model.

To detect problems by on board staff and passengers The estimated rate of non-detection of a hot axle box by passengers and by on-board staff (OBS) is represented by equation 3.

$$RDH_{OBS} = (RDH_{PA})^{NP \times PIQ} + RDH_{CPA} \tag{3}$$

- Assuming a low passenger attention: $RDH_{PA} = 5.10^{-1}$ / h.
- Taking into account a high speed train with all seats taken (377 seats) and 8 wagons: $NP = 377 \div 8 = 47$.
- Assuming that 25% of passengers will want to alert a member of the crew and 12,5% (1 in 8 cars) will find one: $PIQ = 0,25 \times 0,125 = 0,03125$.
- Assuming a low procedural action by the crew: $RDH_{CPA} = 5.10^{-2}$ / h.

The result is $RDH_{OBS} = 4,1.10^{-1}$ / h.

To monitor axle boxes by STEM agents The estimated rate of non-detection of a hot axle box by STEM agents is represented by equation 4.

$$RDH_{STEM} = \left(\frac{RDH_{STEMA}}{SAQ}\right)^{NA} \tag{4}$$

- Assuming an average attention of STEM agents to traffic safety: $RDH_{STEMA} = 5.10^{-2}$ / h.
- Assuming that 20% of the work of the STEM agents is devoted to traffic safety: $SAQ = 0,20$.
- Assuming that there is, at least, one agent available: $NA = 1$.

The result is $RDH_{STEM} = 2,5.10^{-1}$ / h.

To monitor axle boxes by on track HBD The estimated rate of non-detection of a hot axle box for high speed trains with on track HBD is represented by equation 5.

$$RDH_{HBD-Track} = \frac{NFT}{Factor} \times \frac{1}{(1 - I_{0-HBD}) \times h_{DBC} \times NB_{HBD}} \tag{5}$$

- There is currently $NB_{HBD} = 106$ on track HBD on the French network, it can be estimated that there is a HBD every 35,49 kms.
- Number of failures tolerated per year (NFT) = 1.
- A risk aversion factor of HBD for high-speed lines: Factor = 15.
- Unavailability of HBD (I_{0-HBD}) = $3,96.10^{-3}$.
- Hours of operation: $h_{DBC} = 18 \times 333 = 5994$ h/year (18 hours per day during 333 days per year).

The result is $RDH_{HBD-Track} = 1,05.10^{-7}$ / h.

4.2 Application on a high speed line

The previous SADT model is applied for assessing the hot box protection system for high speed lines. As every sub-functions and activities of the model are supposed independent, the probabilistic evaluation of the whole model for only on-track HBD corresponds to equations 6 and 7.

$$RDH_{model} = RDH_{TCD} \times RDH_{Driver} \times RDH_{OBS} \times RDH_{STEM} \times A \quad (6)$$

$$with\ A = \frac{(STEM - step)^2}{V} \times RDH_{HBD-Track} \times \frac{(HBD - step)^2}{V} \quad (7)$$

- RDH_{TCD} and RDH_{Driver} are neglected.
- The average train speed (V) is supposed equal to 300 km/h.
- The length of the line is supposed to be 1125 kms long with a distance of 90 kms between two STEM railway stations (*STEM-step*) and two HBD are separated from 30 kms (*HBD-step*).

The number of not detected hot boxes is represented by equation 8, it corresponds to the probability rate of the model multiplied by the high speed train traffic in France (121,1 million tr.km).

$$N_{not-detecting} = RDH_{model} \times Traffic \quad (8)$$

Table 1. Results for high speed lines

Rates	Complete system	System without STEM agents	Units
RDH_{model}	$4,82.10^{-12}$	$5,72.10^{-12}$	per hour
RDH_{model}	$1,27.10^{-8}$	$1,51.10^{-8}$	per year
RDH_{model}	$6,73.10^{-3}$	$7,55.10^{-3}$	per mil. tr.km
$N_{not-detecting}$	0,77	0,82	per year

Final results obtained are shown on table 1. One year of use for a high speed train corresponds to 8 hours of use per day during 330 days so 2640 hours of use. This table shows that, thanks to the protection systems in place (on track and staff), the risk of not detecting a hot box is low: 0,77 hot box per year and is similar to frequencies in Europe and U.K. presented in the first part of this paper. Moreover, this risk must be tempered by the probability of having an accident. Indeed, the risk of not detecting a hot box does not necessary conduct to a broken axle because of speed variation of the train and/or of train stops at railway stations. Moreover, table 1 shows that STEM agents detection is limited. Indeed, the not detection rate for the complete system is $6,37.10^{-3}$ per million tr.km. The same rate without STEM agents is $7,55.10^{-3}$. This performance could be due to the limited detection of *visible* hot boxes.

5 Conclusion

In this paper, a method for modeling and assessing railway protection systems is presented. The innovative feature of this method is to establish safety levels without using existing rules. Our method provides risks rates that can be use to validate admittance requirements of infrastructure managers. As an illustration, the hot box protection system is evaluated. Future works will concern the enhancement of the model by adding new functions like *To manage the stop of a train* and to apply this method for freight lines.

References

1. NF EN 50126: Applications ferroviaires : spécification et demonstration de la fiabilité, de la disponibilité, de la maintenabilité et de la sécurité (FDMS). In: Comité Europeen de Normalisation Electrotechnique. UTE, Union Technique de l'Electricite et de la communication, (2000)
2. Gerdun V., SedMak T., Sinkovec V., Kovse I., Cene B.: Failures of bearings and axles in railway freight wagons. In: Engineering Failure Analysis 14: 884–894, (2007)
3. European Railway Agency: The Railway Safety Performance in the European Union, (2009)
4. Benyon J.A., Watson A.S.: The use of Monte-Carlo analysis to increase axle inspection interval. In: Proceedings of the 13th International Wheelset Congress: Rome, Italy, (2001)
5. Smith A.: Fatigue of railway axles: a classic problem revisited. In: Proceedings of 13th European Conference on Fracture (ECF): San Sebastian, 173–81, Spain, (2000)
6. Réglement S2C: Circulation des trains. SNCF, (2006)
7. NF EN 15437-1: Applications ferroviaires : Surveillances des boîtes d'essieux exigences liées aux interfaces - Partie 1. In: Comité Europeen de Normalisation Electrotechnique. UTE, Union Technique de l'Electricite et de la communication, (2009)
8. IGL Technology: SADT : un langage pour communiquer. Eyrolles, Paris, ISBN 2212081855, (1989)
9. Vilmart C.: Bilan LOTI du contrôle de vitesse par balises (KVB). RFF, (2008)
10. Van Elslande P.: Dynamique des connaissances, catégorisation et attentes dans une conduite humaine située. L'exemple des *erreurs accidentelles* en conduite automobile. PhD Thesis, Paris V René Descartes University, (2001)

Estimation of Safety Requirements for Wayside Hot Box Detection Systems

Sonja-Lara Bepperling[1] and Andreas Schöbel[2]

[1] ETH Zürich, Institut für Verkehrsplanung und Transportsysteme,
Wolfgang-Pauli-Str. 15, CH-8093 Zürich, Switzerland
bepperling@ivt.baug.ethz.ch
[2] Vienna University of Technology, Intitute of Transportation,
Karlsplatz 13/230-2, A-1040 Vienna
andreas.schoebel@tuwien.ac.at

Abstract. The estimation of safety requirements for wayside train monitoring systems is becoming more and more relevant due to the fact that new technologies are currently developed to recognize various fault states in railway operation during vehicles run. If such systems try to enter the market soon, the question arises in the process of accreditation at the national notified body. To support this process the BP risk methodology was used in this paper to estimate the required safety integrity level for such monitoring devices.

Keywords: Risk Analysis, Hot Box Detection Systems, BP-Risk

1 Introduction

For the past twenty years, hot box detectors and fixed brake detectors are used by most European railway infrastructure companies to prevent derailments [5], [10], [11], [13], [14], [15]. As a proxy for the risky situation, the temperature of the axle bearing and the brake disc are measured. As infrared sensors are easily available, there are many different manufacturers. The varieties can be found in the sensor geometry used for measuring and the consequential ascertainable types of bogie constructions.

Typically, two hot box detection sensors, one provided on each side of the track, measure the axle boxes. Simultaneously, a further infrared sensor measures the temperature of the brake disc for fixed brake detection and a hot wheel detection sensor measures the temperature of the wheel flange to detect critical temperatures of blocked brakes. Axle counters are mostly used for fixation of axles at hot box detection sensors, for calculation of the sum of axles passing and for fixation. Visualization of the results from measurements is often done by a customary PC with WINDOWS operating system. Moreover, all data transmitted from track-side equipment, can be stored in a centre and if necessary exchanged with other systems.

E. Schnieder, G. Tarnai (eds.), *FORMS/FORMAT 2010*,
DOI 10.1007/978-3-642-14261-1_14, © Springer-Verlag Berlin Heidelberg 2011

In case of an alarm, one infrastructure manager has to inform the driver of a train that a wayside hot box detection system has recognized a temperature exceeding a warning limit. It is also possible to declare two limits of temperature for warning and for triggering an alarm. In both cases an inspection of the axle is needed. This will be done by technical inspectors where available or by the driver of the train. Important for the braking process is the normal use of brake power and to prevent an emergency brake because the forces involved could cause a derailment.

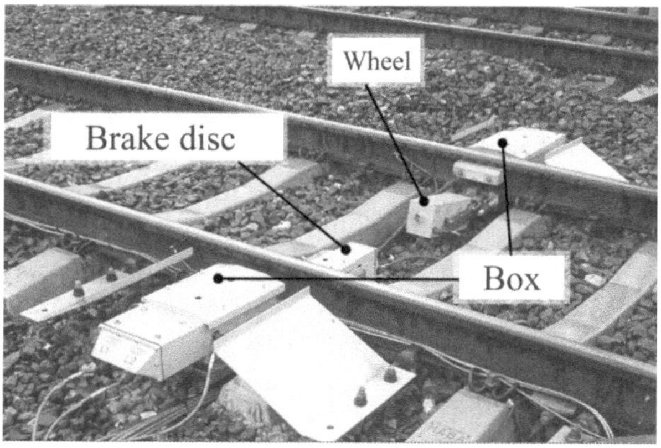

Fig. 1. Hot box and hot brake detection system used by Austrian Railways (ÖBB)

Staff (train driver, station inspector or wagon examiner) can only ascertain visually if an axle journal is broken, an axle-bearing is glowing, or an axle-box case is deformed. Even if none of these indicators can be found, the train will continue its journey with reduced maximum speed to the next place where a technical inspector is available. Otherwise - if the driver verifies the defect - the wagon has to be removed even when the alarm has been generated by the locomotive itself.

Because of the early development of hot box detectors national standards were mainly used for their design. In accordance to the current standards of CENELEC [3] the question arises if such systems have to fulfill certain safety requirements (by means of a tolerable hazard rate and safety integrity level). For a rough estimation of the recommended safety requirement, the so called Best Practice Risk (BP-Risk) analysis offers a suitable methodology to answer this question.

2 Application of BP-Risk for Hot Box Detection

2.1 BP-Risk

BP-Risk is a semi-quantitative approach for railway risk assessments, which has been published [2] and validated [1].

Semi-quantitative methods are a combination of qualitative and quantitative approaches. In [9], they are defined as "qualitative, model-based" risk assessment methods. This means, that for semi-quantitative risk assessment methods, numerical (quantitative) values are assigned to qualitative scales.

Examples for semi-quantitative risk methods can be found in the automobile industry and in the IEC 62061 [8] standard "Safety of machinery".

For BP-Risk semi-quantitative implies, that on the outside, the risk analyst uses the front-end tables, provided by BP-Risk to assess the risk parameters. On the inside, there exists a risk model, which is implemented in the tables and actually uses numerical input values. Therefore BP-Risk uses the following risk model:

$$R = f \cdot g \cdot s \tag{1}$$

where f is the hazard frequency - expressed as a Tolerable Hazard Rate (THR), g is the probability, that the considered hazard leads to an accident, and s represents the potential damage.

The two risk parameters g and s are divided into sub parameters to ease their assessment.

The general approach for risk assessments with the help of BP-Risk includes the common aspects (as required by standards and regulations):

- System definition
- Hazard identification
- Consequence analysis with BP-Risk tables
- THR derivation with BP-Risk table

These steps are described in the following for the application of BP-Risk for the safety requirements of hot box detection systems. It has to be noted, that this is not a complete risk assessment, but that this paper tries to highlight the most important aspects and tries to bring out some first reasonable results.

2.2 System definition

Hot box detectors can be interpreted as monitoring devices to ensure the track guiding by measuring temperature as an indicator for failures at boxes or brakes. Thus the hot box detectors have to recognize if the temperature is exceeding a specified threshold. The considered function could be called: "hot box detection".

For this paper, the following assumptions were made:

- As 90 till 95% of all hot boxes are observed on freight trains, only freight traffic is considered for this analysis. The average speed limit for most European freight trains is 100 km/h.
- We assume, that an on-site staff (e.g. station inspector) watches the trains and thus is able to discover possible hot boxes. Therefore it might be possible to prevent a hazard or even an accident through human mitigation.
- Experts estimate that every fifth to tenth hot box eventually leads to a derailment.

2.3 Hazard identification

If the hot box detection fails unnoticed and a hot box accurse, the resulting event could be called "track guidance not ensured". One possible consequence could be a derailment caused by a broken axle stub (Fig. 2.). Thus, the considered hazard is the "failure of the hot box detection". Note that the hazard scenario assumes that a hot box already exists.

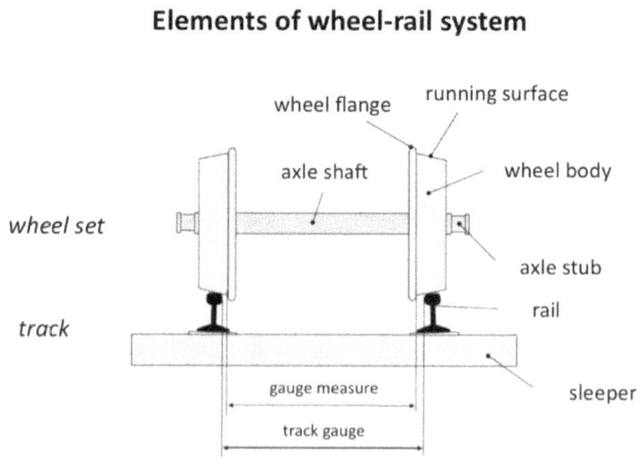

Fig. 2. Elements of wheel-rail system (based on [12])

Another possible consequence of a hot box could be a fire in the vehicle. For now, this accident type is excluded within this analysis to reduce complexity.

2.4 Consequence analysis

To assess the potential mitigation of an accident and the possible consequences of an accident, the BP-Risk parameters are used for the risk analysis.

To assess parameter G, which assesses possible mitigation factors, two sub parameters are used: parameter B and M.

Subparameter B was previously called “operating density”, because it considers the possibility of a train entering an occupied track. For hot box detection and in particular for derailments, this aspect is not crucial. Thus, the original meaning of the quantitative parameter B was used in this case, which is the “confrontation probability of disadvantageous circumstances”, defined by [7]. This confrontation probability implies a situation in which at least one counter measure exists that can prevent the impending loss. Here, this probability is considered as how likely a hot box leads to a derailment. The experts judgment of every fifth to tenth hot box leading to a derailment is interpreted as a rarely. Thus, for parameter B, value 2 is chosen (refer to Table 1.). Note, that we assumed, that a hot box already exists. Thus the mitigation factor of not having a hot box, when the hot box detection fails is not considered here and has to be included in a following causal analysis.

Table 1. Probability of confrontation (Parameter B)

B	probability of confrontation	explanation
1	low	hardly ever does the hazard lead to an accident
2	regular	rarely does the hazard lead to an accident
3	high	frequently does the hazard lead to an accident

Subparameter M assesses if human mitigation is possible. Thus, it assesses a situation where the hazard already exists and where only human intervention can prevent an accident. For hot box detection this could be the train driver or an on-site staff member (e.g. a station inspector). The train driver has no possibility to detect a hot box by himself - sometimes he doesn't even recognize a derailment when only one wheel derails and the air pipe is still working well. Therefore the only potential intervention can be carried out by an on-site staff member who could observe a hot box and then prevent the train form driving on (e.g. by immediate stopping of the train in terms of interlocking system). We assume that this can be considered as a rule-based action, because it is not daily routine, but still possible. Hence, the M value is determined to be 3 for the on-site staff member and 5 for the train driver, if you refer to Table 2. Therefore we can take 4 the value for parameter M.

Accordingly, parameter G has the following value: $G = B + M = 2 + 4 = 6$

To assess the potential damage S, three sub parameters are used: parameter T, V and A.

Subparameter T considers the mass of the trains, because parameter S takes the kinetic energy into account. As Table 3. illustrated, the more mass the trains have, the higher is the T-value and thus the higher will be the

Table 2. Human prevention (Parameter M)

M	human prevention	explanation
1	often possible	"skill-based" action under disadvantageous circumstances
3	seldom possible	"rule-based" action under disadvantageous circumstances
5	almost never possible	random human intervention

risk value afterwards. As we consider a derailment, the mass of the trains doesn't really play a role for this accident type. It can even be advantageous to prevent a derailment, if the train is heavier because of the higher Q-force in relation to the Y-force (in accordance with the derailment criteria defined by NADAL). Therefore, we won't consider subparameter T and thus assign it the value 0.

Table 3. Train category (Parameter T)

T	train category	example
1	short-distance passenger traffic	local train, rapid-transit, commuter rail
2	long-distance passenger traffic + high speed traffic	trainset, passenger train, night train, motorail train
3	freight traffic	freight trains

The decisive speed V is estimated to be around 100 km/h for our considered freight trains. This would correspond to a high speed when referring to Table 4. Thus, the value for V is chosen to be 3.

Table 4. Decisive Speed (Parameter V)

V	decisive speed	example
1	minor	shunting, running at sight, freight corridor
2	medium	line with limited traffic
3	high	local line, regional service
4	very high	long distance or high speed line

Subparameter A assesses how many people might be affected by the potential accident. In our example, we consider a derailment as our typical accident type. But we also consider freight trains, where there are no passengers on board the train.

Usually derailments of a freight train don´t lead to harm of persons but only to damage to property. Hence, we can estimate the number of affected people to be a no person but only damage to property, which is not considered by BP-Risk at this stage. That corresponds to A having the value of 0.

But it has to be noted, that follow up events like collisions or even fire of dangerous goods are not considered here, because those would be worst case scenarios which have a very low frequency of occurrence. BP-Risk uses typical scenarios, which lead to typical consequences. If worst case scenarios would be assessed, there is no need for a method like BP-Risk, because one would have to derive the highest safety requirements anyway which lead unnecessary high costs for the components.

Table 5. Number of affected persons (Parameter A)

A	number of affected persons	example
1	single person	collision with obstacle (not other train)
2	few persons	collision at level crossing
3	several persons	derailment
4	many persons	
5	very many persons	head-on or end-on collision (of trains)

Parameter S can then be calculated by adding the three subparameters: S = T + V + A = 0 + 3 + 0 = 3. Altogether the sum of G and S is 6 + 3 = 9.

3 THR Derivation

To derive the THR for the considered function, Table 6. is used. The sum of G and S corresponds to a certain tolerable hazard rate (THR). For our example, this would be THR $= 3 \cdot 10^{-5}/h$. Note, that BP-Risk uses RAC-TS as a risk acceptance criterion which was implement in Table 6.

It is important to note, that the hot box detection is not determined to a technical component only at this stage. The function itself can be carried out by on-site staff only or by technical components or a combination of the two although there would be differences in reliability and availability. Moreover networked technical components would allow trend analysis and therefore a higher level of operational safety and performance. In case of networked components, an in depth analysis would have to take into account that the devices are "connected" with each other.

Also an in depth analysis would have to consider the aspect of having only technical wayside components. For this example, we assumed as a simplifica-

Table 6. Table for deriving a tolerable hazard rate (THR)

$THR = (\sqrt{10})^F$ (perfunction)	G + S	Description
$3 \cdot 10^{-5}/h$	9	once in 3 years
$10^{-5}/h$	10	once in 10 years
$3 \cdot 10^{-6}/h$	11	once in 30 years
$10^{-6}/h$	12	once in 100 years
$3 \cdot 10^{-7}/h$	13	once in 300 years
$10^{-7}/h$	14	once in 1,000 years
$3 \cdot 10^{-8}/h$	15	once in 3,000 years
$10^{-8}/h$	16	once in 10,000 years

tion, that the devices are independent from each other and don´t belong to a network.

4 Conclusion

The application of BP-Risk for the derivation of safety requirements for hot box detection systems shows that such devices can be designed with a THR of around $3 \cdot 10^{-5}/h$ per function. It was necessary to adapt some of the well designed parameters within the BP risk procedure to fit also for this application. BP-Risk offers the possibility to receive a rough estimation for the safety requirements of hot box detection systems within a short time. For each parameter the arguments considered in the decision making process are comprehensible and allow therefore a simple modification in case of updates.

Further steps would be to carry out an in depth analysis for certification of hot box detection systems. The results of this paper shall be considered as a first approach for a possible way forward.

References

1. Bepperling, S.-L.: Validation of a semi-quantitative approach for railway risk assessments. PhD thesis, Institute of railway systems engineering and traffic safety, Technical University of Braunschweig (2008). (This thesis is only available in German: http://www.digibib.tu-bs.de/?docid=00024255).
2. Braband, J.: Risikoanalysen in der Eisenbahn-Automatisierung. In: Eurailpress Edition Signal + Draht, Hestra-Verlag, Hamburg (2005)
3. CENELEC: Railway application - Communications, signalling and processing systems - safety related electronic systems for signalling, EN 50129. (2003)
4. Deutsche Bahn AG: DB Richtlinie 413, Bahnbetrieb- Infrastruktur gestalten. valid since 01.01.2002, Version of 2006
5. Eisenbrand, E.: PHOENIX MB: Die neue Dimension in der Heißläuferortung. In: Signal + Draht 93 (2001)

6. European Parliament: Commission Regulation (EC) No 352/2009 of 24 April 2009 on the adoption of a common safety method on risk evaluation and assessment as referred to in Article 6(3)(a) of Directive 2004/49/EC of the European Parliament and of the Council. EN L 108/4 Official Journal of the European Union, 29.4.2009
7. Hinzen, A.: The influence of the human error on the safety of railways. Dissertation, Institute of Transport Science and Chair of Railway Engineering and Transport Economics, Institute of Technology of North Rhine-Westphalia RWTH Aachen (1993)
8. IEC: Safety of machinery - Functional safety of safety-related electrical electronic and programmable electronic control systems. IEC 62061 (2005)
9. Milius, B.: A new classification for risk assessment methods. In: Proceedings FORMS/FORMAT 2007. Braunschweig, Hrsg. Schnieder, E. und Tarnai, G.: Formal methods for Automation and Safety in Railway and Automotive Systems, pp. 258 - 267
10. Mironov, A., Tagirov, A.: Use of complexes KTSM in modern conditions. Automation, Communications, Information Science 9 (2002) p. 59
11. Mironov, A.: "New potentialities of KTSM and ASK PS", Automation, Communications, Information Science 12 (2005) p. 64-67.
12. Pachl, J.: "railroad construction", lecture notes, Institute of railway systems engineering and traffic safety, Technical University of Braunschweig, 2003.
13. Rottensteiner, U.: "VAE-HOA 400 DS - Heißläuferortungsanlagen für finnische Hochgeschwindigkeitsstrecken", Signal + Draht 95 (2003) p. 6-10.
14. Schöbel, A., Pisek, M., Karner, J.: "Hot box detection systems as a part of automated train observation in Austria", EURNEX - ZEL 2006, Zilina; 30.05.2006 - 31.05.2006; in: "Towards the competitive rail systems in europe", (2006), 8080705518; p. 157 - 161.
15. Šhvalov, D., Šapovalov, V. "Systems for diagnosis of rolling-stock: textbook for technical schools and colleges of railway transport", edited by D. Švalov. Moscow; "Maršrut", (2005); 268 p.

Formal Specification and Automated Verification of Safety-Critical Requirements of a Railway Vehicle with Frama-C/Jessie

Kerstin Hartig[1], Jens Gerlach[1], Juan Soto[1], and Jürgen Busse[2]

[1] Fraunhofer FIRST, Berlin, Germany
{kerstin.hartig, jens.gerlach, juan.soto}@first.fraunhofer.de
[2] Institute of Railway Technology, IfB GmbH, Berlin, Germany
jb@bahntechnik.de

Abstract. Formal verification of software provides a higher level of assurance than classical software testing. In this paper, we report on our experience with the Frama-C/Jessie verification tool in the railway domain. We analyse safety-critical requirements of a railway vehicle, formalize them using the ANSI/ISO-C Specification Language (ACSL) and achieve automated proofs to verify that the implementation satisfies the formal specification. The main requirement for the successful application of Frama-C in the railway domain is its qualification according to EN 50128.

Keywords: Frama-C/Jessie, ACSL, Unit Proof, EN 50128, Railway Domain

1 Introduction

Developers of industrial critical embedded software face numerous challenges. They must ensure their software provides high levels of assurance, often, within tight time constraints. In addition, they generally have limited budgets for conducting software testing. However, regardless of how much software testing is performed, one cannot fully guarantee that a safety feature is satisfied. Hence, one must resort to the use of formal methods. Notwithstanding, it is the assessor and the certification authority that must be convinced that software meets the safety-critical requirements contained within standards, such as the EN 50128 standard for railway software.

Our work addresses the promotion and adoption by industry of a particular formal method, namely, deductive verification, based on an open source tool, called Frama-C within the framework of the DEVICE-SOFT[3] project.

[3] DEVICE-SOFT stands for Deductive Verification of Industrial Critical Embedded Software.

E. Schnieder, G. Tarnai (eds.), *FORMS/FORMAT 2010*,
DOI 10.1007/978-3-642-14261-1_15, © Springer-Verlag Berlin Heidelberg 2011

Frama-C is a suite of tools dedicated to the analysis of the source code of software written in C. Frama-C is particularly interesting for us because C is an important programming language for the development of safety-critical embedded systems. The development of Frama-C is coordinated by CEA LIST[1] in Saclay, France. The Frama-C platform is extensible, meaning users can incorporate new plug-ins. Jessie[2] is a plug-in which enables deductive verification of C programs within Frama-C. It is based on the computation of weakest preconditions. Given an ACSL[3] annotated C program, the Jessie plug-in generates intermediate code and verification conditions for use by automated theorem provers or interactive proof assistants.

Airbus has successfully used Caveat (a predecessor of Frama-C/Jessie) to replace unit tests by by unit proofs in parts of their software[4]. With a vast array of specialty domains to choose from, we opted to focus on the railway industry, given the fact that it is one of the domains with a vested interest in the usage of formal methods[5]. Another aspect of the DEVICE-SOFT project is the development of a deductively verified C-library of standard algorithms[6].

The particular safety requirements for the vigilance device of a generic diesel locomotive (see Section 2) were drawn from an experienced assessor in the evaluation of railway software. Starting from these requirements we follow the simplified V-model depicted in Figure 1.

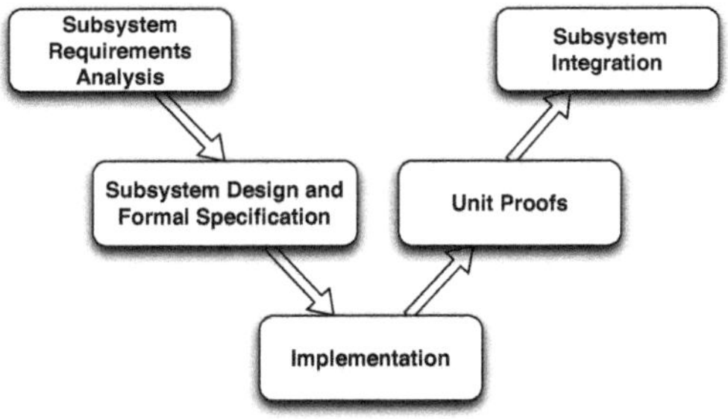

Fig. 1. Formal Subsystem Development Process.

Initially, we analyse the informal requirements of the vigilance device (see Section 2). We formalize these requirements using the ANSI/ISO C Specification Language (see Section 3). Thereby we discuss some of the most important features of ACSL. In Section 4, we will illustrate some important aspects that have to be considered when implementing the ACSL specification. Section 5 shows the results of the automated unit proofs that show that the implementation satisfies its ACSL specification. Finally, we discuss the benefits and challenges and the qualification of the software tools.

2 Requirements Analysis

The vigilance device monitors the alertness of a train operator. Such a mechanism must be introduced to detect whether a train operator has fallen asleep, is ill, or possibly has died. While the train is in operation, the driver must prove his/her attentiveness by invoking the vigilance device regularly. e.g., by pushing a button. The informal requirements read: 1.) If the train operator does not apply at least one of the vigilance device pedals or buttons within a timed interval of eight seconds the forced brake must apply automatically. 2.) If the train operator continuously applies at least one of the vigilance device pedals or buttons for more than thirty-five seconds the forced brake must apply automatically.

Figure 2 depicts the UML state diagram for the vigilance device described above.

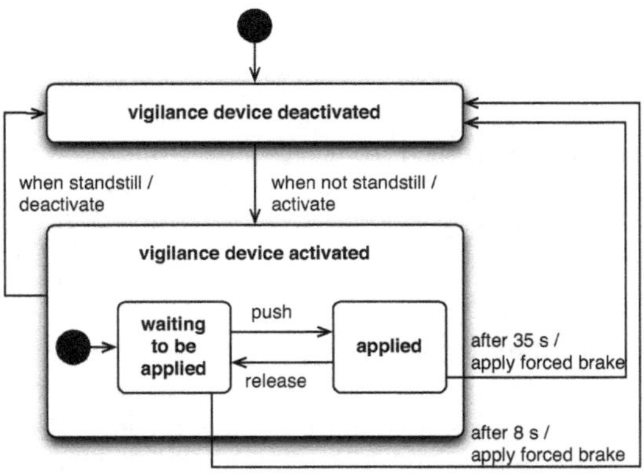

Fig. 2. UML State Diagram of a Vigilance Device.

Note that the state diagram is already more specific than the informal specification since it introduces different states of the vigilance device. With these states one can specify that no buttons need to be pushed while the train is at standstill.

3 Subsystem Design and Formal Specification with ACSL

For our case study we consider user-defined data types containing the core data of the locomotive and the vigilance device. Figure 3 depicts the model of the data structures `Locomotive` and `Vigilance_device`.

In `Locomotive` the variable `actual_time` represents a relative-time counter expressed in milliseconds (msecs). The variable `speed` measures the velocity of the train in km/h. The variables `standstill`, `train_brake` and

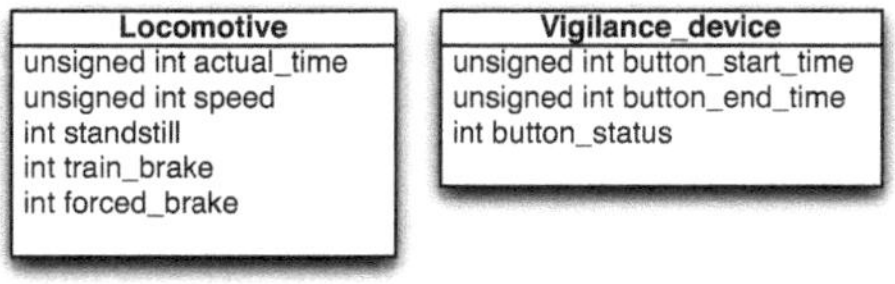

Fig. 3. Data Structures.

`forced_brake` are represented by an integer, but contain the Boolean values true or false, where `1` denotes `TRUE` and `0` denotes `FALSE`.
The structure `Vigilance_device` contains the variables `button_start_time` and `button_end_time` describing the moments a vigilance device pedal or button was invoked and released the last time. The variable `button_status` represents a Boolean variable, which is true if any vigilance device pedal or button is currently invoked.

3.1 An Introduction to ACSL

ACSL annotations are expressed in special C-comments `/*@...*/` as a multi-line comment or `//@...` as a single-line comment. A function contract declares a set of **requires** clauses, stating the properties the function may expect on entry, and a set of **ensures** clauses, stating the properties the function must satisfy upon exit.

Properties are formulas denoted in a language close to `C` itself. For example, equality, negation, and conjunction are denoted by `==`, `!`, and `&&`, respectively; binding-priorities are as in `C`. In addition, the weaker-binding junctors `==>` and `<==>` denote implication and equivalence. Moreover, relation chains familiar from mathematical notation, such as, `0 <= i < n`, may be used.

Function parameters and visible variables may appear in formulas, they refer, by default, to their values on entry and exit in a **requires** and **ensures** clause, respectively. The notation `\at(v,L)` refers to the value of `v` at the program point corresponding to the `C`-label `L:`. A predefined label `Old` allows one to refer to on-entry values in **ensures** clauses too, `\old(EXPR)` being an abbreviation for `\at(EXPR,Old)`. The label `Here` refers to the on-exit value in an **ensures** clause.

A macro-like mechanism, the **predicate** definition allows users to abbreviate arbitrary formulas by a parametrized name. Parameters may be of `C` types or logical types enclosed in parentheses). They may also denote memory states at certain labels enclosed in curly braces.

3.2 Using ACSL

Due to some informal requirements some variables may only take on Boolean values, i.e., `1` represented as `TRUE` and `0` represented as `FALSE`. However, since an integer may take more than these two values we can formalize this restriction by formulating a user-defined predicate with this constraint as follows:

```
//@ predicate true_or_false(int a) = (a == FALSE)||(a == TRUE);
```

The type invariant for `Vigilance_device` can be expressed by the predicate `vigilance_invariant` as shown below. It is required that the status may only be true or false as it's a Boolean value. Therefore, we use our formerly defined predicate `true_or_false`. Furthermore, there exists an equivalence between the status of the buttons and the start and end time of their invocation. If, and only if, the moment of releasing a button or pedal is after it was pushed last time the value of the status is true.

```
/*@
  predicate vigilance_invariant{L}(Vigilance_device* vig) =
    true_or_false(vig->button_status) &&
    (vig->button_status <==> vig->button_end_time < vig->button_start_time);
 */
```

Similarly, we can also formulate a type invariant for the user-defined data type `Locomotive`.

Below we can see an example of a predicate that checks whether the limit for holding a vigilance pedal or button has expired. Due to the informal specification this predicate remains true if any vigilance pedal or button is currently invoked (`status` must be true) and the difference between the actual time and the moment the button invocation was started exceeds thirty-five sec., represented by `MAX_VIGILANCE_BUTTON_HOLD`.

```
/*@
  predicate vig_button_hold_expired{L}
  (Vigilance_device* vig, Locomotive* loc) = vig->button_status &&
    (loc->actual_time - vig->button_start_time > MAX_VIGILANCE_BUTTON_HOLD);
 */
```

Similarly, we can formulate a predicate that checks whether the pause limit between the invocation of any vigilance pedal or button has expired.

The function `process_vigilance_device` checks whether the vigilance device is processed correctly, which means, the vigilance buttons or pedals are neither held nor paused too long. We can see the function contract for such a function below.

```
/*@
  requires \valid(vig) && \valid(loc);
  requires locomotive_invariant(loc) && vigilance_invariant(vig);
  requires actual_time_is_latest_time(vig, loc);

  ensures locomotive_invariant(loc) && vigilance_invariant(vig);
  ensures actual_time_is_latest_time(vig, loc);

  behavior forced_brake_initiate:
    assumes !loc->standstill &&
      (vig_button_hold_expired(vig, loc) || vig_button_break_expired(vig, loc));
    assigns loc->forced_brake && loc->train_brake;
    ensures loc->forced_brake && loc->train_brake;
```

```
  behavior no_forced_brake_necessary:
    assumes loc->standstill || (!loc->standstill &&
        !vig_button_hold_expired(vig, loc) &&
        !vig_button_break_expired(vig, loc));
    assigns \nothing;

  complete behaviors;
  disjoint behaviors;
 */
void process_vigilance_device(Vigilance_device* vig, Locomotive* loc);
```

Since the function contains pointers that must be dereferenceable, we use the `\valid` clause to specify this requirement. We include our user-defined predicates `vigilance_invariant` and `locomotive_invariant` as pre- and postconditions into the function contract in order to emulate type invariants. Additionally, we expect the formerly defined predicate `actual_time_is_latest_time` to hold as pre- and as postcondition, which means no time variable may have values describing the future relative to the actual time.

At the end of that contract we can specify that the described behaviors are `complete` and `disjoint`, as they include all possibilities and both exclude each other. The inclusion of these last clauses provides an additional check on the specification itself.

4 Implementation

Below, the implementation of the function formally specified in the former subsection is depicted. Given the very precise formal specification, the implementation of this function is straightforward.

```
void process_vigilance_device(Vigilance_device* vig, Locomotive* loc) {
  if (!loc->standstill){
    if (check_vig_button_break_expired(vig, loc) {
      loc->train_brake = TRUE;
      loc->forced_brake = TRUE;
    }
    else if (check_vig_button_hold_expired(vig, loc)) {
      loc->train_brake = TRUE;
      loc->forced_brake = TRUE;
    }
  }
}
```

This function calls two auxiliary functions:

```
int check_vig_button_hold_expired(Vigilance_device* vig, Locomotive* loc);

int check_vig_button_break_expired(Vigilance_device* vig, Locomotive* loc);
```

Each of those, of course, must be formally specified as well. This example shows, that the programmer needs to have knowledge about the specification language as well. This process is comparable to the fact, that the programmer must also be able to write unit tests.

5 Results of Deductive Verification

Deductive verification, i.e. *unit proving*, was performed using the Frama-C release Boron[7] and the software verification platform Why 2.26 including Jessie. The tools were executed on Mac OS X (10.5 Leopard). The following five automated theorem provers were referenced in our experiments:
Alt-Ergo[8], CVC3[9], Simplify[10], Yices[11], and Z3[12]. Table 1 depicts the results obtained by deductive verification of the specified C-functions using Frama-C and Jessie.

	Automated Theorem Provers				
Proof Obligations	**Alt-Ergo**	**CVC3**	**Simplify**	**Yices**	**Z3**
Total	74	74	74	74	74
Valid	74	72	74	70	74
Timeout	0	2	0	4	0
Proven	**Yes**	**No**	**Yes**	**No**	**Yes**

Table 1. Results of Deductive Verification

Table 1 shows 74 verification conditions (VC), a.k.a., proof obligations, generated by the verification condition generator, which is called implicitly upon invoking Jessie. In our case study the three provers Alt-Ergo, Simplify and Z3 were able to discharge all verification conditions as valid and thus could prove that the implementation of the functions satisfies their specification. CVC3 and Yices, on the other hand, were not able to prove all conditions in the allocated time frame of 10 seconds and thus reported timeouts.

6 Conclusions and Future Work

We have demonstrated an approach and a procedure for verifying safety critical software components in an automated way. In our experience, an ACSL specification can form a crucial artifact for communicating within the development process because many participants in this process can relate to ACSL, e.g., requirements engineers, architects, software developers or assessors.

Regarding the scalability of our approach, we think that the main focus of deductive verification will be on relatively small, well understood software components and on software subsystems that consist of such components. However, even so the return on investment of deductive verification compared to traditional component testing is not yet well understood and will be part of our future research.

Nevertheless, there are still challenges to manage. First of all, formal specification and deductive verification need to be properly integrated in the software development process. This is of utmost importance for high-integrity applications where authorities must be convinced to accept replacing unit tests by unit proofs.

In order to use deductive verification efficiently, advanced user knowledge is expected. Software developers must be trained in the art of writing annotations and in the use of Frama-C tools. If a proof cannot be accomplished it may be non-trivial to determine the source of the problem. Better feedback (recommendations) must be provided by the tool to aid in troubleshooting.

Safety-critical software in the railway domain must be assessed before it is placed into service. The assessment process is described in the European standard EN 50128:2001. From the assessors point of view, the main requirement for deductive verification methodology and tools is their qualification. Qualification in that context means a particular trustworthiness. That trustworthiness must be proven and documented.

The EN 50128 standard does not address verification tools directly. Requirements in the field are in evolution. To summarize, the main requirements are documentation of the tools behavior and of any constraints in its use, combined with successful validation of the tool. The safety argument for the tool, has to address, that faults within the source code must not be proven. One approach to reach that goal is undertaking test suites on source code with intentionally intruded faults.

Acknowledgments. This work was completed within the DEVICE-SOFT project, which is supported by the Programme Inter Carnot Fraunhofer of the Federal Ministry of Education and Research (BMBF) (Grant 01SF0804) and the Agence Nationale de la Recherche (ANR).

References

1. *CEA LIST, Laboratory of Applied Research on Software-Intensive Technologies*, `http://www-list.cea.fr/gb/index_gb.htm`.
2. Marché, C., Moy, Y.: *Jessie Plugin Tutorial, Boron Version*, `http://frama-c.com/jessie/jessie-tutorial.pdf`, (2010).
3. Baudin, P., Filliâtre, J.-C., Marché, C., Monate, B., Moy, Y., Prevosto, V.: *ACSL: ANSI/ISO C Specification Language, Version 1.4*, http://frama-c.cea.fr/download/acsl_1.4.pdf, (2009).
4. Souyris, J., Favre-Felix, D.: *Proof of properties in avionics*, IFIP Congress Topical Sessions, pages 527-536, (2004).
5. Jim Woodcock, et al.: *Formal Methods: Practice and Experience*, ACM Computing Surveys, Volume 41, Issue 4 (October 2009).
6. Burghardt, J., Gerlach, J., Hartig, K., Soto, J., Weber, C.: *ACSL By Example, Towards a Verified C Standard Library*, `http://www.first.fraunhofer.de/owx_download/acsl-by-example-5_1_0.pdf`, (2010).
7. Correnson, L., Cuoq, P., Puccetti, A., Signoles, J.: *Frama-C User Manual, Boron Release*, `http://frama-c.com/download/user-manual-Boron-20100401.pdf`, (2010).
8. *Alt-Ergo Theorem Prover*, `http://alt-ergo.lri.fr/`.

9. Barrett, C., Tinelli, C.: *CVC3*, In Proceedings of the 19th International Conference on Computer Aided Verification (CAV'07), Volume 4590 of Lecture Notes in Computer Science, pages 298-302, (2007).
10. *Simplify Theorem Prover*, http://secure.ucd.ie/products/opensource/Simplify/.
11. Dutertre, B., de Moura, L.: *The YICES SMT Solver*, http://yices.csl.sri.com/tool-paper.pdf.
12. de Moura, L., Bjørner, N.: *Z3: An Efficient SMT Solver*, Conference on Tools and Algorithms for the Construction and Analysis of Systems (TACAS), Budapest, Hungary, http://research.microsoft.com/projects/z3/z3.pdf, (2008).

Simulation and Optimization of the Longitudinal Dynamics of Parallel Hybrid Railway Vehicles

Maik Leska, Robert Prabel, Andreas Rauh, and Harald Aschemann

Chair of Mechatronics, University of Rostock
Justus-von-Liebig-Weg 6, 18059 Rostock, Germany
{Maik.Leska, Robert.Prabel, Andreas.Rauh, Harald.Aschemann}
@uni-rostock.de

Abstract. In this paper, a basic simulation of the longitudinal dynamics of parallel hybrid railway vehicles is presented, which is used for an optimization of the operating strategy. An internal combustion engine, the primary power source, is supported by an electrical energy storage system with motor and generator allowing for joint usage of both energy sources. For the main components of the power train, interface variables representing power flow and energy, are defined. A combined quadratic performance index is introduced for a system optimization. Here, an optimization of fuel consumption and emissions is considered for a parallel hybrid structure to minimize the chosen performance index.

Keywords: Hybrid Railway Vehicles, Emissions, Optimization, Fuel, Consumption, longitudinal Dynamics

1 Introduction

Not only in automotive applications but also at railway vehicles, hybrid systems have a promising potential to reduce fuel consumption and exhaust gas emissions by essentially improving the energy efficiency. Railway system suppliers worldwide invest in the research and development of such systems. Generally, a hybrid propulsion system uses more than one power source. The choices of energy storage devices, further system components, and the overall system structure depend on the vehicle's duty cycle and other issues such as cost effectiveness and serviceability. The energy savings mainly depend on the duty cycle and on the capacity and time constants of the available energy storage device. For instance, for shunting vehicles, a large peak power is required only for a short time span that is followed by an extended period of idling. This suggests downsizing the conventional engine by using a powerful energy storage device. In passenger applications, the majority of energy savings does not originate from the engine downsizing but rather from energy savings resulting from recuperation of braking energy and its reuse during subsequent

E. Schnieder, G. Tarnai (eds.), *FORMS/FORMAT 2010*,
DOI 10.1007/978-3-642-14261-1_16,

acceleration phases. Compared to automotive applications, the great advantage during the development of operating strategies for hybrid railway vehicles is the fact that the duty cycles are known beforehand. Thereby, optimal operating modes can be computed almost completely offline. Hillmansen and Roberts carried out a kinematic analysis of energy storage systems which suggests a potential of up to 35% energy savings for commuter vehicles [1]. The result of the research work on hybrid concepts for diesel multiple units in [4] suggests a potential for reduction of fuel consumption of up to 25% on a fixed route. In [3], a description of modern onboard storage technologies is given. These are flywheel systems, hydrostatic accumulators, double layer capacitors, and batteries. All predictions of energy saving potentials in the above-mentioned studies arise out of simulation studies, a combination of modeling and optimization is not yet state of the art.

In this paper, a simulation structure for a parallel hybrid railway vehicle is shown in Sec. 2. An exemplary mathematical modeling is given for the vehicle. In Sec. 3, an optimality criterion is formulated by a parameterizable performance index quantifying fuel consumption and emissions. Sec. 4 presents the simulation and optimization results. Finally, Sec. 5 concludes this paper and gives an outlook on future research.

2 Modeling of the Basic Parallel Hybrid Structure

The power train of a basic parallel hybrid railway vehicle mainly consists of an internal combustion engine, an electric motor, and an energy storage device. Fig. 1 shows the schematic block diagram of the parallel hybrid system in which the blocks visualize the models of the hybrid system components. Here, all main effects contributing to the longitudinal dynamics were modeled. To build up a complete system model, consistent interface variables representing the flow of power and energy were defined for each component. Hence, other structures, for instance, a serial hybrid structure, can be built up easily by rearranging the components of the drive chain. In principle, two alternative calculation approaches are possible for simulation studies of hybrid power trains: either the forward or the backward calculation. The dashed arrows in Fig. 1 stand for the forward calculation, where the main input is the engine torque commanded by the driver. The remaining system variables are determined by numerical integration as indicated by the dashed arrows. Note that these arrows represent the order of the calculation steps and not the direction of power flows.

The backward approach corresponding to the solid arrows, on the contrary, represents the solution to an inverse problem. Here, all system variables are derived from the duty cycle subject to the chosen operating strategy.

In the optimization, only the backward approach is employed. For a given duty cycle, the necessary drive torque is calculated in the vehicle subsystem. The resulting power is split after the planetary gear box between the electric motor/generator and the internal combustion engine. The output is given

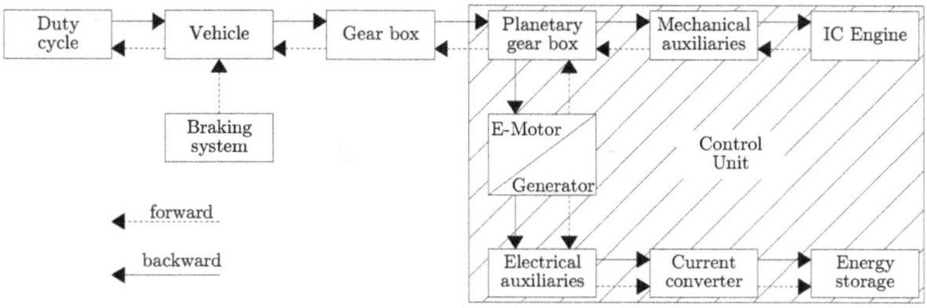

Fig. 1. Simulation structure for a parallel hybrid railway vehicle.

by the torque which is necessary to follow the duty cycle. Additionally, fuel consumption, NOx- and PM emissions are determined. The way how the electrical part (consisting of the electric motor/generator, the energy storage device, and the electrical auxiliaries) is used, depends on the operating strategy to be optimized in Sec. 3.

2.1 Forward and Backward Calculation on the Vehicle

In Fig. 2, the block of the vehicle subsystem and its interface variables are depicted. The different kinds of arrows again indicate the direction of the forward and the backward calculation, respectively. The inputs of the forward

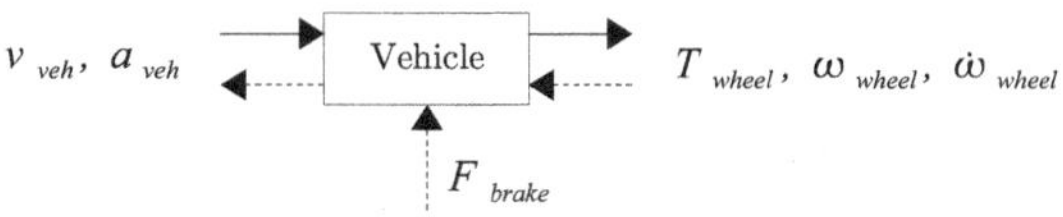

Fig. 2. Interface variables of the vehicle model.

calculation match the outputs of the backward calculation and vice versa. The braking force F_{brake}, however, is considered as an additional driver input for the forward approach only. The mathematical model in this paper is restricted to the longitudinal dynamics because the lateral dynamics do not have an important influence on the optimization task to be performed. In the forward calculation, the corresponding force balance can be stated as

$$m_{veh}\, a_{veh} = F_{wheel} - F_{brake} - F_{res} - F_{inc} \ . \tag{1}$$

Here, m_{veh} represents the total vehicle mass. The force F_{wheel} acting at the wheels is related to the torque T_{wheel} at the wheel and the wheel diameter d_{wheel} according to

$$F_{wheel} = \frac{2\, T_{wheel}}{d_{wheel}} \ . \tag{2}$$

The resistance force F_{res} describes the friction forces at the wheels as well as the air resistance. It can be approximated by a polynomial

$$F_{res} = k_0 + k_1\, v_{veh} + k_2\, m_{veh} + k_3\, v_{veh}\, m_{veh} + k_4\, v_{veh}^2 + k_5\, v_{veh}^2\, m_{veh} \ , \quad (3)$$

where v_{veh} is the vehicle velocity. With the inclination angle γ, the inclination force F_{inc} is given by $F_{inc} = m_{veh}\, g\, \sin(\gamma)$. Then, the outputs of the forward mode, a_{veh} and v_{veh}, follow from numerical integration.

For the backward approach, the inverse problem is considered using the given duty cycle parameters, namely, the velocity profile v_{veh} and the acceleration profile a_{veh}. Solving the equation of motion (1) for F_{wheel} leads to

$$F_{wheel} = m_{veh}\, a_{veh} + F_{res} + F_{inc} \ . \quad (4)$$

Here, the braking force F_{brake} is included in F_{wheel}. The resistance force F_{res} and the inclination force F_{inc} can be calculated for the given duty cycle. The outputs of the backward approach, the power variables, result in

$$T_{wheel} = F_{wheel} \frac{d_{wheel}}{2}, \quad \omega_{wheel} = \frac{2 v_{veh}}{d_{wheel}}, \quad \dot{\omega}_{wheel} = \frac{2 a_{veh}}{d_{wheel}} \ . \quad (5)$$

2.2 Hybrid and Non-Hybrid operating Modes

Like above, for all system components depicted in Fig. 1, mathematical models were derived in form of algebraic or differential equations as well as quasi-static maps. The control unit determines the power distribution between the combustion engine and electrical energy storage in Fig. 1. Besides the energy management, the operating strategy has to be determined by the control unit. In this paper, the chosen overall operating strategy involves six different modes.

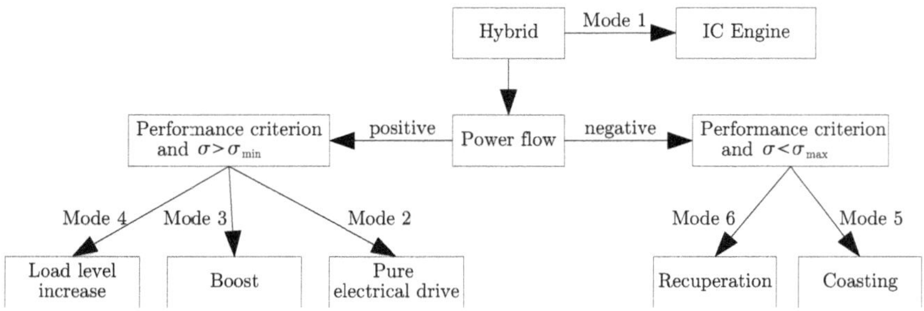

Fig. 3. Hybrid drive strategy.

Mode 1 represents the pure engine mode, in which the internal combustion engine provides the total power demand. Mode 2 is the pure electric mode, thus, only the electric motor is active. In the Mode 3, the power boost mode, both the electric motor and the internal combustion engine operate simultaneously. In the load level increase (Mode 4), the operating point of the combustion engine provides more power than required for following the defined duty cycle. The excess power is used for recharging the energy buffer. This mode is also used at stillstand to fully recharge the energy storage. In Mode 5, the coasting mode, neither the electric motor nor the combustion engine are running. Finally, in Mode 6, the recuperation mode, kinetic energy of the vehicle is recovered in deceleration phases. In Mode 1 up to Mode 4, the power is provided by the combustion engine or the electric motor, that means the power flow is positive. During braking or coasting the power flow changes its direction such that the power is provided by the kinetic energy of the vehicle. The choice between the operating modes is influenced by the duty cycle and the state of charge σ of the energy storage device. In Fig. 3, a simple structure for selecting the corresponding operating mode is depicted. It is shown that according to the direction of the power flow, the choice of the operating mode depends on the performance index and the state of charge σ. The performance index reflects the operating strategy, for example, a minimization of both fuel consumption and emissions, see Sec. 3.

3 Parametrization of Optimality Criteria

At optimization, usually a reduction of the operating costs is aimed at. However, low emissions are of increasing importance as well. Hence, a combined performance index is introduced, and a corresponding optimized operating strategy is derived. Figure 4(a) shows a typical fuel consumption map of a diesel engine. The best specific fuel consumption b_{fuel} is around 60% of ω_{max} and 50% of P_{max}. Longer operations in the partial-load area should be avoided, for example, using Mode 2 at the beginning of acceleration phases or Mode 4 to charge the battery. Besides saving fuel, a reduction of the specific fuel consumption decreases CO_2 emissions as well. CO_2 originates from the combustion of hydrocarbons. The combustion also causes production of other reactants like nitrogen oxides NOx and particulate matter PM. The major origin for nitrogen oxides is the generation of thermal NO, espescially at high temperatures. With increasing loads the temperatures rise as well, and therewith, the generation of nitrogen oxides, see also Fig. 5(b) in Sec. 4. The main component of PM is unburned carbon. The highest PM emissions take place at high velocities with low loads, see Fig. 5(c). Another cost factor may arise from battery ageing, which can be influenced by the admissible depth of discharge or by the average charging and discharging rates. Battery ageing will be addressed in future publications.

To take into account both fuel consumption and emissions, a combined performance index is defined as the weighted sum of the squared normalized components according to

$$J = \int_0^T \left[\alpha \left(\frac{\dot{V}(t)}{V_{nom}} \right)^2 + \beta \left(\frac{\dot{N}(t)}{N_{nom}} \right)^2 + \gamma \left(\frac{\dot{C}(t)}{C_{nom}} \right)^2 \right] dt + \Delta R \ . \quad (6)$$

In (6), the instantaneous fuel consumption is denoted by $\dot{V}(t)$, the NOx and PM emissions by $\dot{N}(t)$ and $\dot{C}(t)$, respectively. The nominal values V_{nom}, N_{nom} and C_{nom} represent the total fuel consumption and emissions respectively, for a non-hybrid vehicle with the given duty cycle in Fig. 4(b). The factors α, β and $\gamma > 0$ are used to weight the three influence factors in the combined performance index.

Obviously, the operating mode that minimizes the performance index corresponds to a purely electrical strategy, in which the complete traction power is provided by the battery. Therefore, the recovered energy by recuperative braking has to be sufficiently large to recharge the battery in the ideal case. The state of charge at the final destination has to be equal to the state of charge at the start to provide a fair comparison of hybrid and non-hybrid operating modes. For that reason, the energy storage device will be recharged at the final destination if there are deviations from the initial value $\sigma(0) = 100\%$ using Mode 4. The emissions and fuel consumption while recharging are considered in the performance index by $\Delta R > 0$.

Constraints mostly concern the physical operating limits, like the maximum engine torque, the motor power and the state of charge [2]. Other constraints are related to the velocity profile and time table, for instance the latest possible time of arrival at the final destination. There are two alternatives to influence the fuel consumption and the emissions. The first is to vary the duty cycle by keeping the bounds on the velocity and final time. The second alternative is to change the operating strategy. In this paper, only the second alternative is considered with a fixed duty cycle.

4 Simulation and Optimization Results

The backward simulation structure for a parallel hybrid railway vehicle in Fig. 1 was implemented in Matlab/ Simulink. For all simulations, the same duty cycle is used, see Fig. 4(b). The output variables can be chosen as the specific and absolute values for fuel consumption and NOx/PM emissions. To obtain a reference for optimization, a non-hybrid strategy was simulated.

4.1 Heuristic Hybridization Strategy

In this section, a heuristic operating strategy for the parallel hybrid, is considered see Fig. 4(b). Here, the normalized velocity is depicted over time. The duty cycle is split into several parts by vertical dash-dotted lines. The numbers in brackets above the trajectory stand for the preselected operating mode in each part according to Sec. 2. Recharging at the final destination is considered by Mode 4. In acceleration phases, the diesel engine is supported by the electrical motor, characterized by the intensity of boost x. This variable, which quantifies the percentage of the electrical contribution to the total traction power, represents the single optimization variable. In parts with a constant velocity, load level increase according to Mode 4 is used to recharge the energy storage. The coasting phases in Mode 5 are applied to save fuel and to minimize the emissions. Unfortunately, coasting can be only utilized if the timetable allows for extended travel times. During the whole duty cycle auxiliaries are considered, which leads to a positive minimum power request to the diesel engine.

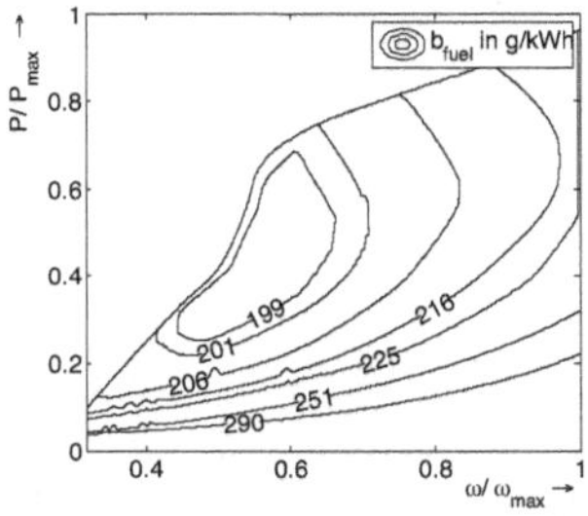

(a) Fuel consumption map.

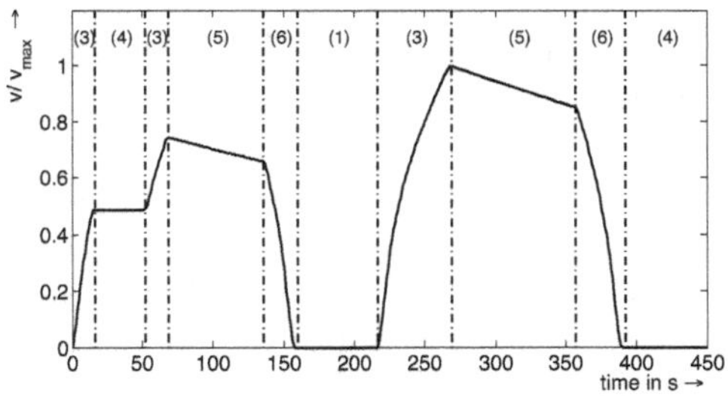

(b) duty cycle with operating modes.

Fig. 4. Fuel consumption map for a diesel engine and fixed duty cycle.

4.2 Optimized Operating Strategy

An improved strategy can be found by minimizing the performance index (6). For that purpose the fminsearch-optimizer of the optimization toolbox in Matlab is used, which is based on the Nelder-Mead method. As single optimization variable, the intensity of the boost x is used. In every iteration one complete simulation is performed including both the given duty cycle as well as the recharging of the energy storage at the final destination. To determine the global minimum, multiple starting points are employed. In

Table 1, the optimization results for different weighting factors α, β, and γ are stated. The values V, N, and C stand for the overall fuel consumption and emissions, these are calculated within the simulation. To obtain the same dimensions all values are standardized to the reference values of the non-hybrid strategy. Depending on the chosen weights, the strategy can be

Table 1. Optimization results.

α	β	γ	x	J	V/V_{nom}	N/N_{nom}	C/C_{nom}
1	0	0	25.2%	0.7018	0.8378	0.9608	1.0291
0	1	0	21.6%	0.9162	0.8451	0.9572	1.0224
0	0	1	$<$ 0.01%	1.0000	1.0181	1.0000	1.0000
1	1	1	21.5%	2.6691	0.8424	0.9573	1.0213

switched to minimal fuel consumption/emissions. For example, minimal fuel consumption is reached with an intensity of boost of $x = 25.2\%$, the minimal NOx emissions are reached with $x = 21.6\%$, and the minimal PM emissions are reached with $x < 0.01\%$, which means a purely diesel vehicle. The simulation results show a possible reduction of the fuel consumption by 16% ($V = 84\%$) compared to a pure diesel vehicle. The emissions could not be reduced drastically, only a marginal reduction of the NOx emissions could be reached. The PM emissions increase by 2% compared to a pure diesel vehicle. The reasons are the long phase of recharging compared to the driving time and the operating strategy. The boost mode is activated in each acceleration phase independent of the velocity. This causes a large increase in the specific PM emissions at high velocities, see Fig. 5(c). These rises can be avoided if velocity constraints are introduced for Mode 3.

5 Conclusions and Outlook on Future Research

In this paper, a mathematical modeling for a basic parallel hybrid structure, with calculation performed in backward direction, has been presented. The simulations were performed for a parallel hybrid structure with a specified duty cycle. By minimization of an appropriate performance index, an optimized operating strategy was found. The simulation and optimization results show a reduction of fuel consumption of about 17% for a fixed duty cycle.

In future research, the optimization shall be extended to a larger number of optimization parameters. Besides the optimization of the operating strategy, the duty cycle will be optimized as well. Furthermore, alternative hybrid structures are to be investigated.

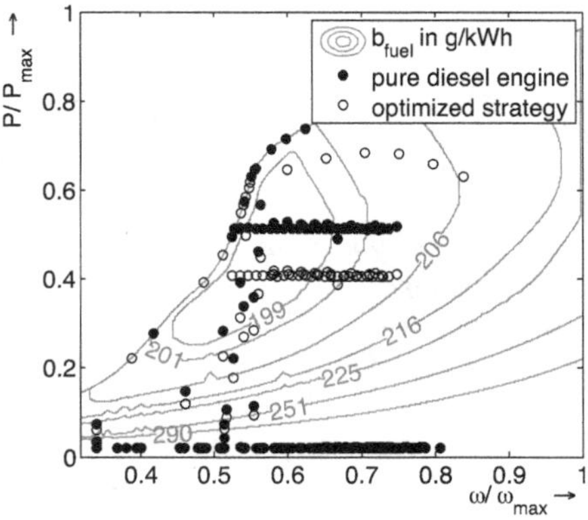

(a) Fuel consumption.

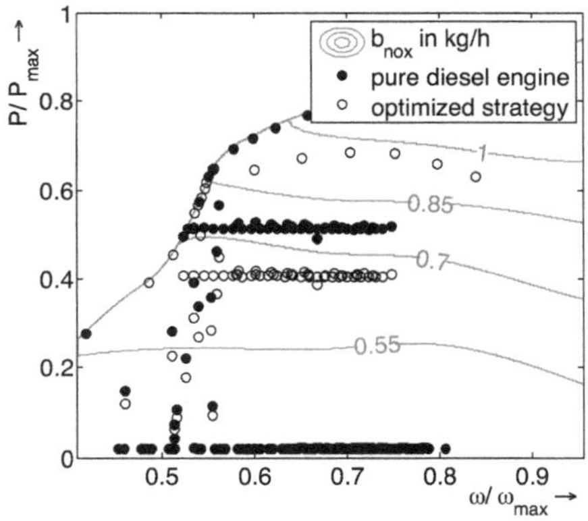

(b) *NOx* emissions.

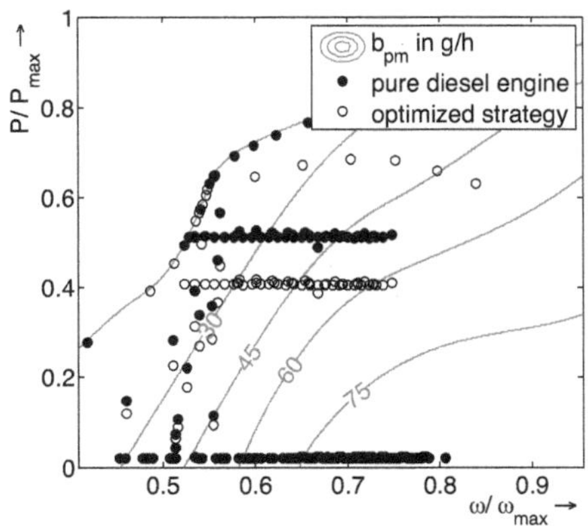

(c) *PM* emissions.

Fig. 5. operating points in the fuel consumption map and in the emissions maps.

References

1. Hillmansen, S., Roberts, C.: Energy Storage Devices in Hybrid Railway Vehicles: a Kinematic Analysis. Proceedings of the Institution of Mechanical Engineers, Part F: Journal of Rail and Rapid Transit **221** 135–143 (2007)
2. Guzzella, L.: Vehicle Propulsion Systems, 2nd edition, Springer, Berlin Heidelberg (2007)
3. Lu, S., Meegahawatte, D.H., Guo, S., Hillmansen, S., Roberts, C., Goodman, C.J.: Analysis of Energy Storage Devices in Hybrid Railway Vehicles, Proc. Intl. Confenrence on Railway Engineering, Hong Kong, China, 2008
4. Hillmansen, S., Roberts, C., McGordon, A., Jennings, P.: Draft Final Report: DMU Hybrid Concept, Evaluation - Follow on Work, DfTRG/0078/2007, 2009, `http://www.railway.bham.ac.uk/documents/Hybrid_Rail_Report_DMU_V1.pdf`

2nd Day Sessions

3rd December 2010

- Legal Framework and Risk Metrics
- Automotive Assistance Systems, Development and Simulation
- PLC Verification and Test Tool Chain

Dissemination of the Commission Regulation (EC) No 352/2009/EC on Common Safety Method on Risk Evaluation and Assessment

Maria Antova, Dragan Jovicic, and Thierry Breyne

European Railway Agency, Safety Unit, Safety Assessment Sector,
Rue Marc Lefrancq 120, F-59307 Valenciennes Cedex, France
{maria.antova, dragan.jovicic, thierry.breyne}@era.europa.eu

Abstract. In order to support the market opening across Europe, the European Commission decided to define a common and harmonised approach for managing the railway safety. To take this forward, the EU Legislators have approved in April 2004 the railway safety directive 2004/49/EC. This directive allocates amongst others the task of defining a Common Safety Method (CSM) on risk evaluation and assessment to the European Railway Agency (ERA). The Regulation 352/2009/EC covering this CSM on risk assessment was published in April 2009 in the EC official journal. In order to support the railway actors in the implementation of this Regulation, as well as in order to gain inputs for its upcoming revision, the European Railway Agency performed a series of dissemination workshops for the CSM on risk assessment. The objective of this paper is to summarise and highlight some points from the dissemination of the Common Safety Method on risk assessment.

Keywords: EC, Regulation, Common Safety Method (CSM), Risk Assessment, Dissemination, Questions, European Railway Agency (ERA).

1 Introduction

The safety directive 2004/49/EC [1] in its Article 6(3)(a) requires the development of a harmonised approach for risk assessment. This development led to a Commission Regulation on a Common Safety Method on risk evaluation and assessment (EC) N° 352/2009 [2] published in the Official Journal of the European Union on 29th of April 2009. The risk assessment and risk management processes, which it describes, contain basically the following three steps:

1. identification of hazards;
2. risk analysis and risk evaluation based on existing risk acceptance principles, identification of safety measures and resulting safety requirements;
3. demonstration of the system compliance with the identified safety requirements.

E. Schnieder, G. Tarnai (eds.), *FORMS/FORMAT 2010*,
DOI 10.1007/978-3-642-14261-1_17, © Springer-Verlag Berlin Heidelberg 2011

Additional requirements for mutual recognition were included in the process: Hazard Management and independent assessment of the correct application and results of the CSM process performed by an Assessment Body. The process has to be applied to any safety related change of the railway system in a Member State, which is considered to be significant.

Due to the relative novelty of some aspects of the formal CSM process for risk assessment, this CSM Regulation will have a gradual implementation. From 19 July 2010, it applies:

- to all significant technical changes affecting vehicles as defined in Article 2(c) of Directive 2008/57/EC [3];
- to all significant changes concerning structural sub-systems, where required by Article 15(1) of Directive 2008/57/EC or by a TSI.

In order to give sufficient time to the concerned actors, where needed, to learn and apply the new common approach as well as to gain experience from it, the CSM Regulation remains voluntary with respect to the operational or organisational changes until 1 July 2012.

For the purpose of providing further support and detailed presentation on the CSM Regulation to all the actors from the European railway community, in the period of mid-2009 until mid-2010, the Agency has organised a series of CSM dissemination workshops. This dissemination exercise is also considered as a part of the activity related to the fulfilment of the requirement in Article 9(3) of the CSM Regulation. It requests ERA to monitor and collect feedback on the application of the CSM on risk assessment in order to make recommendations to the Commission with a view to improving it. The dissemination workshops were supported by a questionnaire, which was filled in by the participants of the workshop in advance. The purpose was also to allow the Agency to see, at an early stage, how the railway actors understand the concepts defined in the CSM Regulation to adapt the presentation to their needs and to collect feedback, ideas and suggestions how to improve the CSM in the scope of its upcoming revision.

2 Topics from the dissemination of the CSM on risk assessment and evaluation

In the following sections there is a presentation of some questions, which have been discussed during the CSM dissemination workshops, which took place so far. It is to be paid attention that it is not possible to cover all the questions, within the scope of one single paper. All topics, which have been discussed during this series of dissemination meetings, are considered to be equally important.

The gained inputs are going to be considered in the scope of the revision of the CSM on risk assessment, the work on which is going to start at the beginning of 2011. The report for the revision of the CSM on risk assessment should be made by ERA and sent to the Commission by the end of 2011.

2.1 How should the requirements from the CSM Regulation be fulfilled

In order to enable the mutual recognition of the results of risk assessments and to ensure that the existing safety levels are maintained in the Community rail system, the CSM harmonises the process for risk assessment. It specifies *what* requirements must be fulfilled *without* specifying *how* to fulfil them.

This is an important property of the CSM, since the current practices for risk management vary between the different Member States. Usually, different tools are established and different practices for the assessment and evaluation of risks exist. Moreover, the specific physical and historical properties of the railways around Europe request different activities in their given context.

Therefore, it is necessary for the proposers to be able to comply with the CSM requirements, paying attention to their local particularities and adjusting their activities to the particular context of their work. In their activities, this freedom and flexibility are considered to be very important.

2.2 Significant change

The CSM on risk assessment shall apply to any safety related change of the railway system in a Member State, which is considered to be significant.

If there are no notified national rules defining whether a change is significant or not in a Member State, the proposer shall decide, by expert judgement, on the significance of the change. This decision is based on the following criteria that are provided in Article 4 of the CSM on risk assessment:

- failure consequence;
- novelty used in implementing the change;
- complexity;
- monitoring;
- reversibility of the change.

If the change is not significant, the CSM application is not mandatory, but the decision needs to be documented. This would allow the national safety authority to check it during its supervisions of the proposer's safety management system.

For non significant changes, attention on the possible additionallity effects needs to be paid. This means that the expert judgement shall always evaluate, if when added up, the sum of all non significant changes since the last application of the CSM becomes a significant change.

In the scope of the dissemination of the CSM, discussions took place, on the question, if it is viable to provide a harmonised list of all changes in the railway system, which are considered to be significant. On one hand such a checklist would avoid the need to count on expert judgement when deciding on the significance of a change and would thus bring some strict rules to the answer of this question. On the other hand, having such a predefined list in a harmonised piece of European legislation would, by a way, take away the responsibility of the proposer to decide if, based on its experience, the change is significant and would also have the effect that every time when a change is not published on the list, a Europe-wide solution of this problem will need to be sought. Further questions exist, on:

- is the development of such a list feasible;
- how can such a list take into account the different operators' expertise;
- would not it affect the cost of risk assessments for those railway actors who could have decided that the change is not significant;
- is it going to influence positively or negatively the process for answering to the question of the significance of a change and the demonstration of CSM compliance?

During the work on the development of the CSM Regulation, the solution to support the decision by the predefined criteria referred to above has been put in place. This seems to be a good compromise between setting up a too restrictive and inflexible framework or setting a question with a too wide spectrum of answers, without any further guidance on how to proceed with its answer. In the scope of the CSM dissemination workshops, with the help of the pre-workshop questionnaire, it has been confirmed that the criteria are used already nowadays by the railway actors in order to define whether a change, which they would like to introduce, is significant. Already before the CSM Regulation has come into force, the actors are answering to such questions, in order to define the way how to analyse their changes and the resulting hazards, in the scope of their safety management systems. They are used to applying this practice in their everyday work and are confident in knowing well its effectiveness level.

2.3 Hazard Record

The CSM Regulation requires that the proposer in charge of the risk management process maintains a hazard record. The hazard record is the document in which identified hazards, their related measures, their origin and the reference to the organisation, which has to manage them, are recorded and referenced.

It is clear that the hazard record is an important part of the hazard management process. It helps to document and support the decision making process, by providing transparency and consistency. Due its traceability, on one hand it allows corrective actions to be taken promptly and quickly, and on the other hand, it supports the exchange of information between the different actors. It gives them the possibility to contribute to the evidence of a continuous compliance with the relevant requirements. It does not need to be complicated, because it is mainly targeted on the key issues.

Whereas these advantages of the hazard records are obvious and clear for most of the railway actors, among them there are also such, who admit that they are not yet experienced in its usage. Therefore, advice on different questions often seems to be needed.

For example, for some actors it is not very obvious to know when the hazard record has to be updated. This is usually done whenever:

- a new hazard is discovered or a new safety measure is identified during the design phase;
- a new hazard is identified during the operation and maintenance of the system after its commissioning, so that the hazard can be assessed in compliance with the CSM as to whether it represents a significant change;
- it could be necessary to take into account accident and incident data;
- there are changes to the safety requirements or the assumptions about the system;

Additionally, it is good that a difference is made between the hazard records, which are mainly meant to be used in the preventive part of the railway safety, and the statistics from railway accidents and incidents, which every company is maintaining and is a typical reflection of the reactive part of the railway safety. Although, the prevention and the reaction are interconnected with each other, still, typically the hazard record is structured and maintained in a different way from the event statistics of the railway companies. Nevertheless, it is normal that it receives input for its updates, whenever such unwanted events occur, which lead to the recognition of new hazards, or to the change of some of the existing ones.

Another topic of particular interest is the maintenance of the hazard record in the typical situation when there are a number of actors involved in a certain significant change, each of whom has to have responsibility for his part of the system under assessment. In such cases, it is normal that each of the involved parties keeps a record of the hazards for their part of the assessed change or project. Usually there is one overall actor (proposer) who has responsibility for the main record. This main record covers all the necessary elements of the system under assessment. It does not have to contain all the information from the other actors, but it needs to keep the links and key safety related issues. The exchange of information gains more importance, if the hazard cannot be controlled by one actor alone.

The practice has shown that when classifying the hazards in the hazard records, every company is having its own logic. Even though, at the very beginning it might be a bit challenging to set up such a logic, which is adjusted to the company needs and is able to reflect its existing safety profile, once it has been set, it becomes a powerful support for the creation and the maintenance of the hazard records. Sometimes it might be hard to figure out the necessary level of detail for the documentation of the hazards. Nevertheless, by gaining more practice in the usage of the hazard record, the companies are gradually learning, which level of detail is most proper for their situation too.

Currently, there is no requirement in the CSM Regulation on the question how long to maintain the data from the hazard records. Therefore, also this decision lies in the responsibility range of the proposers.

2.4 Tools and Methods for risk assessment

During the CSM dissemination workshops, questions have been asked about when to use which risk assessment tools and where could more information on this topic be found.

Some of the European railway actors have requested further support and more detailed training on how the risk management and risk assessment process described in the CSM Regulation can be applied. For some part of them, the risk management and assessment concepts, as well as the respective terminology of a risk based approach seem to be new. Sometimes, even if the whole concepts are not new for them in theory, they confirm to be rather inexperienced in their practical implementation. It has been reported that it is often hard to find literature on these topics available in the own language and this implies an additional difficulty for the concerned actors.

Therefore, in order to assist them, the European Railway Agency has decided to elaborate with the support of a subcontractor a training material on the risk management and risk assessment techniques and tools.

For this purpose, a tender for a study has been launched in April 2010. The work on the study has started in August 2010. The estimated date for the completion of the activities lays in February 2011.

The aimed training material shall be supported by an analysis of advantages and disadvantages of the methods and tools used in the different steps of the CSM process. The risk management and risk assessment terminology, the different techniques, tools and methods that might be needed to demonstrate the compliance with the CSM Regulation, will have to be explained. Their use shall be illustrated with instructive examples of risk assessment, which are taken from different spheres of the work in the railway sector (construction of new lines, changes of existing lines, introduction of new and/or modified technical systems, operational changes, etc.).

The developed materials are going to be made available on the web page of ERA. Their exact dissemination modes are going to be proposed and defined in the scope of the work on their development.

2.5 Coordination between ERA and CENELEC

Currently, the CENELEC CLC TC9X Working Group 14 (WG 14) for safety-related standards is working on a combined revision of the following standards:

- EN 50126:1999 [4], which specifies the Reliability, Availability, Maintainability and Safety (RAMS) process for the railway system;
- EN 50128:2001 [5], which concentrates on the software in the signalling;
- EN 50129:2003 [6], concerning the safety relevant electronic signalling systems;
- EN 50155:2001 [7], which is about the electronic equipment used on rolling stock and does not have any special safety or software focus.

From the perspective of the railway community, the revision of these standards must reflect the European safety activities currently undertaken by non-standardisation bodies, one of which is ERA. For this reason, the new standards should provide to the users clear links with ERA's railway safety orientated perspective and activities.

Unlike the CSM on risk assessment, which is a mandatory element for the EU Member States, the compliance with the CLC standards is not mandatory. Nevertheless, they focus on the technical process and outcome in connection with the safety assurance of products and systems both across the EU and beyond. They are going to reflect among others also the risk acceptance principles and criteria, which are currently being developed as a part of the upcoming revision of the CSM Regulation.

The CENELEC standards and the EU regulations are complementary and collectively address broader aspects of safety assurance across the EU. Therefore, having recognised all these aspects of the current developments of both CENELEC and the ERA, a common active coordination activity has been set up between them. Its aim is to align the EN standards with the activities of the Agency and especially with the CSM on risk assessment and its revision, and to provide further a sound and safe basis for the European railways.

3 Conclusions and outlook

Shortly after the publication of the CSM Regulation in April 2009, the European Railway Agency has started a series of activities for its dissemination.

The first step in the CSM dissemination was represented by a series of workshops, which the Agency has held for all concerned railway actors around Europe. During them and with the support of the associated CSM guides [8], [9], the requirements of the CSM Regulation have been explained further and in more details.

With the support of railway sector organisations and national safety authorities, the Agency is going to continue the dissemination activity with the review of the feedback based on real case examples of changes to the railway system whereby the CSM process has been applied. Among others, this should allow to learn from different actors more about the gained experience and the encountered difficulties by the application of the CSM. In this way, these could be taken into account for the future revision of the CSM.

During the dissemination workshops, series of questions have been raised. Most of them were reflecting difficulties, which arise from the particular legal framework of the different Member States and are therefore not proper for a publication in a generalised paper. Nevertheless, there were also questions at a higher level, which have been asked repeatedly and whereby more guidance is necessary. Some of them have been summarised in the sections above. Others, like for example the questions, which consider the Independent Assessment Body for the application of the CSM process, and the questions about the

Risk Acceptance Criteria, which are to be applied for the aims of achieving mutual recognition whenever the third risk acceptance principle (the explicit risk estimation) is applied, are still under development, and are therefore going to be clarified at a later moment in time.

At this stage, it is important to underline that in order to help the actors from railway sector for the application of the CSM on risk assessment, the ERA has also issued the following two informative and not legally binding documents:

- Guide for the application of the Commission Regulation on CSM on risk assessment [8];
- Collection of examples of risk assessments and some possible tools supporting the CSM [9].

These two documents are translated in all EU languages where Member States operate railway and have been made available on the web site of the Agency. Thus, they are meant to represent the first emergency help for the railway actors, who encounter difficulties in the application of the CSM Regulation.

References

1. Directive 2004/49/EC of the European Parliament and of the Council of 29 April 2004 on safety on the Community's railways and amending Council Directive 95/18/EC on the licensing of railway undertakings and Directive 2001/14/EC on the allocation of railway infrastructure capacity and the levying of charges for the use of railway infrastructure and safety certification (Railway Safety Directive)
2. Commission Regulation (EC) No 352/2009 of 24 April 2009 on the adoption of a common safety method on risk evaluation and assessment as referred to in Article 6(3)(a) of Directive 2004/49/EC of the European Parliament and of the Council
3. Directive 2008/57/EC of the European Parliament and of the Council of 17 June 2008 on the interoperability of the rail system within the Community
4. EN 50126:1999 - The Specification and Demonstration of Reliability, Availability, Maintainability and Safety (RAMS)
5. EN 50128:2001 - Communications, Signalling and Processing Systems - Software for Railway Control and Protection Systems
6. EN 50129:2003 Railway Applications - Communication, signalling and processing systems - Safety Related Electronic Systems for Signalling
7. EN 50155:2001 Railway Applications - Electronic equipment used on rolling stock
8. Guide for the application of the Commission Regulation on the adoption of a common safety method on risk evaluation and assessment as referred to in Article 6(3)(a) of the Railway Safety Directive, Version 1.1. from 06/01/2009
9. Collection of examples of risk assessments and of some possible tools supporting the CSM Regulation, Version 1.1 from 06/01/2009

Designing a semi-quantitative risk graph

Birgit Milius

Institute for Railway Systems Engineering and Traffic Safety
Technische Universität Braunschweig
Pockelsstraße 3, 38106 Braunschweig
b.milius@tu-braunschweig.de

Abstract. Usually, risk assessments are done qualitatively or quantitatively. Neither of these two approaches is satisfactory. In the future, risk assessment methods combining the advantages of both approaches are recommendable. Based on the life cycle process as described in DIN EN 50126, the paper discusses guidelines for the construction of semi-quantitative risk assessment methods and presents a newly developed semi-quantitative risk graph for the application to the railway sector.

Keywords: Risk Assessment, Safety, Risk Graph

1 Introduction

For changed or new parts of the railway system a risk analysis has to be performed. In the past, usually quantitative risk analysis methods were used which are difficult due to not enough data and often time consuming.

A possibility for more effective risk analyses is the use of semi-quantitative methods, e.g. risk graphs. Semi-quantitative methods are designed based on a comprehensible mathematical model. In semi-quantitative methods, several parameters are used for the description of risk. For each parameter a scale with discrete classes is constructed from which the suitable parameter classes for the assessment is chosen. From the combination of the selected parameter classes the requirement class and/or the quantitative safety requirement will result.

The presented paper discusses guidelines for the construction of semi-quantitative risk assessment methods and presents a semi-quantitative risk graph for the application to the railway sector.

2 Classification of Risk Assessment Methods

Risk assessment methods are used for years. According to the classification developed in [1] methods can be classified according to their parameter assessment, mathematical foundation and tolerability/risk acceptance argument.

E. Schnieder, G. Tarnai (eds.), *FORMS/FORMAT 2010*,
DOI 10.1007/978-3-642-14261-1_18,

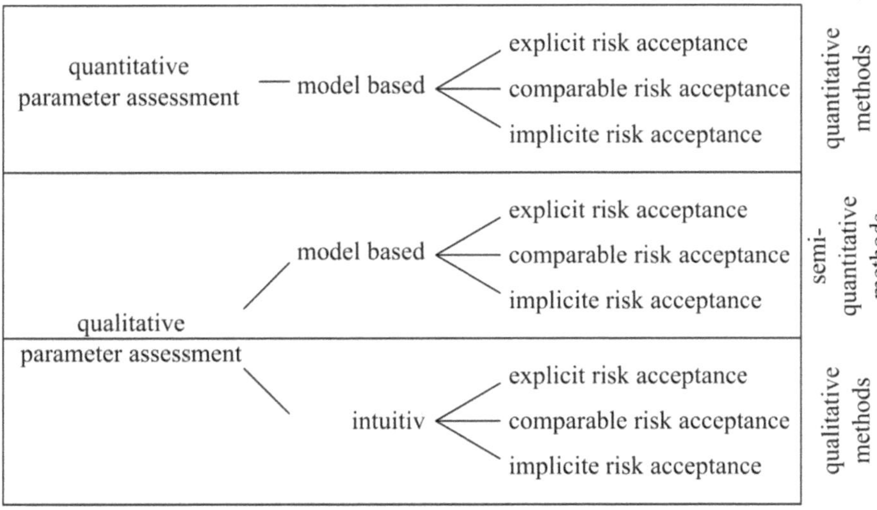

Fig. 1. Risk assessment classification

This has resulted in classifying risk assessment methods as quantitative, semi-quantitative and qualitative (figure 1).

In the past, mainly quantitative and qualitative methods were used.

Quantitative methods are time consuming and therefore rather expensive. They aim at a detailed description of scenarios describing how a hazard can become an accident and looking at all the barriers in between. After the process is described qualitatively using e.g. event trees, petri nets or markov chains, quantification has to be done. Unfortunately, in general no data, not enough or no statistically valid data is known. Therefore, analyses relay heavily on expert's opinion.

Qualitative methods on the other hand are fairly easy to apply. They do not pretend to result in very exact results but make it obvious that the results are rather assessed than calculated. However, it is not proven that the results are significantly different from the results one would obtain with quantitative methods. A major drawback is that qualitative methods were developed over the years using expert's opinion. Therefore, it cannot be proven that the results are actually correct if compared to a tolerable risk.

For the future, hybrid methods, which allow combining the advantages of qualitative and quantitative risk assessment methods, should be used. These methods are called semi-quantitative. They are based on a mathematical model, which allows for a detailed verification of the results and the conditions taken into account. However, these methods are presented in a qualitative way which allows for an easy and fast assessment.

The presented guidelines for the construction process can be applied to any semi-quantitative risk assessment method. The following discussion will lead to the construction of a semi-quantitative risk graph. The risk graph is a method given as example in IEC 61508 [2]. It is referenced in other norms as well. For Nichtbundeseigene Eisenbahnen (Nichtbundeseigene Eisenbahnen are privatly owned / non federal railways.), there is a standard [3] which contains a risk graph actually used for the derivation of safety requirements.

3 Construction Process

3.1 Motivation

The construction process was developed in [4]. It is based on the life cycle process as described in DIN EN 50126 [5]. The main idea is that by interpreting the requirements of each life cycle phase in the light of the construction process, a systematic development process is possible. The life cycle process was developed over the years by outstanding professionals and is part of a norm. One can therefore assume that the life cycle process itself is a complete and very well constructed guidance. The relevant phases of the life cycle process for the construction of a risk assessment method are System Definition and Application Conditions, Risk Analysis and System Requirements.

In the following chapters these three relevant phases are discussed, identifying major aspects to be dealt with when constructing a risk assessment method.

3.2 System Definition and Application Conditions

The *application* sector describes the environment in which the risk assessment method is used. By describing it in detail basic information about actors in and around the system are given.

The description of the application sector is used as a basis for the *system model.* The system model describes how the actors interact. By describing each actor in detail all information needed for the construction of the risk model and consequently the parameters should be given.

The system model itself gives no information about the *analysis level.* However, the analysis level has to be described concisely so that the user knows for which functions the risk assessment can be used. When the user chooses an analysis level different from the one used for the construction process the obtained results are probably not correct.

3.3 Risk Analysis

The risk analysis looks at the functions of a system and derives hazards. To allow for a systematic analysis the *terms hazard and function need to be defined.* This is especially true for the term hazard as the definitions given in the different standards are rather imprecise. The definition of the term hazard should give detailed information e.g. about its relation to an accident and if a hazard is an event or a condition. This is necessary to allow for an exact modeling of the assessed hazard and its risk.

The suitability of a risk assessment method for a certain analysis might depend on the *kind of the result* as this can limit the possibilities of further working with the obtained results. Results of the assessment might be e.g. risk classes or hazard rates.

The qualitative risk model should be developed based on the system model, taking into account the conditions arising from the definition of hazard and function. The qualitative risk model has to incorporate all information which is relevant for a description of risk. A quantitative risk model takes the relevant information from the qualitative model and connects them to calculate a result. Even though in *theory qualitative and quantitative risk* model are developed one after another, in reality it seems advisable to develop both parallel. Often, a qualitative risk model is developed based on an existing equation for risk by describing all factors of this equation in detail. It has to be made sure that the equation is suitable and sufficient for the given assessment as otherwise important influences might be forgotten.

It has to be decided which of the *parameters* of the risk model are dynamic and need to be assessed in the risk assessment process and which parameters are constant. Constant parameters are assessed when constructing the method and will not be changed later. Therefore, it is very important to choose constant parameters with consideration. It might be possible that by choosing certain values for a constant parameter new conditions for the application of the risk assessment method arise.

The dynamic parameters need to be described in classes. To allow for a stringent mathematical derivation of e.g. the final risk class the *parameter classes* should be constructed using the same factor between all classes. By sticking to one factor for all parameter and parameter classes the impact of a parameter class change becomes instantly obvious.

If a constant factor does not seem suitable (e.g. due to largely different value ranges of the parameters) then different factors for the different parameters are possible. To obtain equally spaced results as it is customary for most qualitative risk assessment method today, the derivation process might be done using some rounding. The result will be that the impact of a parameter class change will not be as obvious anymore.

3.4 System Requirements

The designed risk assessment method has to be calibrated to assure that the parameter combinations lead to a correct result. Without a calibration, information about the potential risk of the analyzed function can be obtained, but safety requirements can only be derived when the assessed risk is set in prospective to a benchmark risk. For the calibration of the risk assessment method, a tolerable risk taken e.g. from statistics or given by a regulatory body are suitable. For railways, using the risk acceptance criteria proposed in EU regulation 352/2009 [8] seems to be a good idea.

When the benchmark risk is chosen, a certain parameter combination has to be correlated to the risk. Usually, this can be done by translating the (qualitative) information given with the benchmark risk into parameter classes. If there are no such information then it should be a decision made by experts which accident scenario with which parameter classes describes best the benchmark risk.

When based on the benchmark risk a parameter combination was chosen, the resulting risk classes for other parameter combinations can be derived. This should be done using the quantitative risk model and the factors used for the description of the parameter classes.

4 Example of a Newly Constructed Risk Graph

Application sector: The risk graph was developed as complement to an existing risk graph published in VDV 332 [3]. The VDV risk graph is actually used in practice but upon a closer inspection a lot of flaws are obvious [4]. The new risk graph was developed taking into account the special characteristics of Nichtbundeseigene Eisenbahnen. These are e.g. a reduced maximum speed and typical infrastructural limitations. It focuses on person trains only.

The *system model* (figure 2) was adapted from the system model given in IEC 61508. The main function of a railway system is the transport of persons and goods. Looking at e.g. a single person or a group of persons is difficult. In some analyses the approach was used that by ensuring a safe passage of the trains all persons or goods in the train are also safe. This approach was used by the construction of the system model. The trains are guided by several major functions. Both, the train and the major functions form the railway system. This is enclosed by the environment.

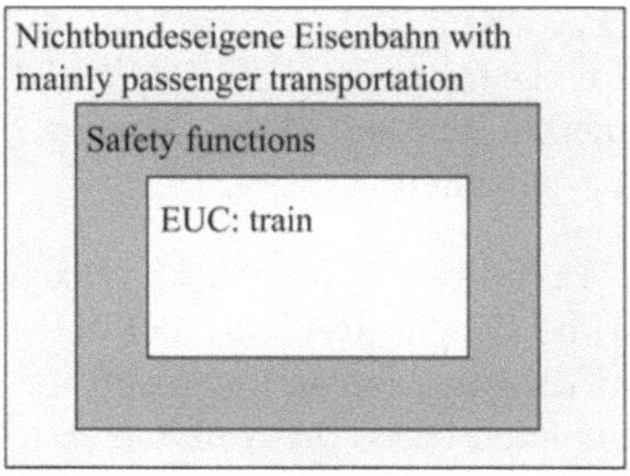

Fig. 2. The system model

Analysis level: According to the VDV-risk graph the risk assessment method shall be used for safety functions. These are all functions which guide a train safely through the railway net. There exist several possible functional descriptions of the railway system which all have advantages and disadvantages. It was defined that the analysis level consists of all functions performed by the physical components which in cooperation with the interlocking or as autonomous components assure the safety of operation. This definition is backed by e.g. the UIC [8]. It has to be shown that the analysis level is the same level for which a benchmark risk is defined.

Definition of function and hazard: The term function was not defined as suitable definitions exist. Hazard was defined as: Hazards in the sense this work are failures to the unsafe side with a potential of relevant harm.

Type of Result: The result of the risk assessment is a hazard rate as it is more flexible than a risk class. A hazard rate can always be transferred into a risk class but a risk class might lead to more stringent results when transferred to a hazard rate.

Qualitative and quantitative risk model: The qualitative risk model is describing in detail the risk to be assessed. Risk is e.g. a combination of severity and frequency. The severity is to be measured as harm to people in the train as this is regarded as a measure of the destructions of the train or its wagons. Another reason for this approach is that German law focuses on personal risk. Therefore, looking at damage to persons allows a clear argument. A collective risk is to be derived as it is assumed that a collective risk for the people in the train is more significant than an individual risk. It was defined that the collective to be looked at consists of passengers and workers but not third parties.

The quantitative risk model for the collective risk was developed based on the equation for individual risk by Braband [6]:

$$CRF = HR \cdot (D + E) \cdot C \cdot F$$

with CRF collective risk of fatality, HR hazard rate, D hazard duration time, E exposure time, C consequence probability and F fatalities. For the quantitative risk model, the equation for CRF was reduced to an equation for risk R

$$R = HR \cdot DE \cdot C \cdot F.$$

Parameters: Three of the four factors DE, C and F of the risk model are used as dynamic parameters for the risk assessment. The fourth factor HR is the result of the risk assessment.

- Harm to people F: An indirect approach was taken were the user chooses the speed at the time of the accident and the type of accident from a diagram and can directly derive the resulting parameter class. The diagrams were constructed based on available accident data (more on the presentation of the parameter severity in e.g. [7]).
- Accident probability C: For derailments, the accident probability was constructed based on the limitations of the infrastructure e.g. relation between the allowed speed and the maximum speed at which a train can pass a switch without derailment. For collisions, the parameter was constructed mainly by combining the rate by which a second train can be encountered and an estimate of the probability for the train to stop on time before another standing or driving train.
- Time of exposure and time of the duration of a hazard DE: For the time of exposure this was calculated taking into account basic information

about the infrastructure (e.g. section length, number of trains per hour). For the duration of the hazard, assumptions have to be made by the user.

All parameters are presented in diagrams allowing for an easy reading. The diagrams make it obvious how changes in the parameter values in one or the other direction on the axis might influence the result. This helps to assess the stability of the results of the risk assessment assuming that the input values usually are estimates and some variation is possible.

Parameter classes: The factor between the parameter classes was chosen to be root of 10. A first choice would have been factor 10 because this factor lies between the safety requirement hazard rates given in e.g. IEC 61508. However, upon closer inspection it became obvious that a factor 10 leads to too wide classes for some of the parameters. Therefore, parameter root of 10 was chosen. As explained earlier, the same factor which is between parameter classes can also be found between the resulting maximum hazard rates.

Calibration: The final risk graph was calibrated using the RAC-TS criteria given in 352/2009 [9]:

For technical systems where a functional failure has credible direct potential for a catastrophic consequence, the associated risk does not have to be reduced further if the rate of that failure is less than or equal to 10^{-9} *per operating hour.*

Taking into account the guidance given in [10], the given benchmark risk is valid on the analysis level.

When calibrating, it was decided that the benchmark scenario given with the EU regulation did not need to be modeled with the most stringent parameter classes. It is supposed that the scenario given with RAC-TS is best described by the parameters F5 (catastrophic, most stringent class), C4 (direct potential, most stringent class), and DE2 (no information given by RAC-TS, chosen based on typical operational conditions). Therefore, the maximum safety requirement which can be obtained with the risk graph is not 10^{-9} failures per operating hour, but 10^{-10} failures per operating hour[1].

Starting with the benchmark combination, by using factor 10 all other hazard rates can be calculated. The final risk graph can be seen in figure 3. The risk graph shows the parameter classes of all three parameters and for each possible combination the resulting hazard rate. Examples for the application of the risk graph can be found in [4].

5 Conclusion

The presented paper shows the main ideas for a lifecycle based construction process of a semi-quantitative risk assessment method. An example of a newly constructed risk graph with the reasoning of its construction process was given.

[1] It is assumed that failure rate as used in [8] corresponds to hazard rate as used in the guidelines.

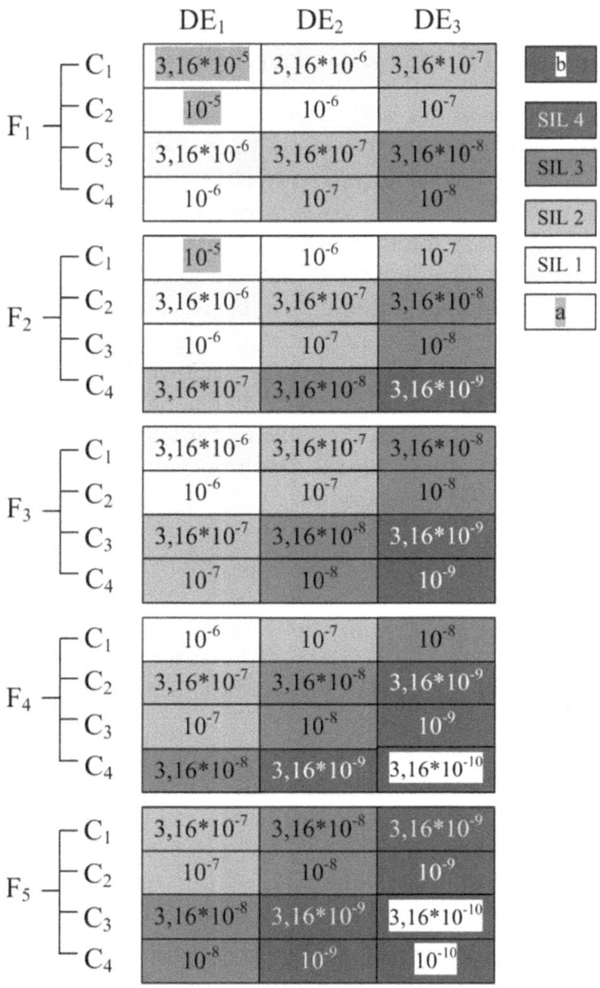

Fig. 3. The newly developed risk graph

The developed guidelines should now be tested and further developed by application. More construction processes of different risk assessment methods ideally in different industries should be documented. The experiences made with the guidelines during the construction process of different assessment methods can be used to define the guidelines.

Furthermore, as semi-quantitative risk assessment methods are rather new, different research areas present themselves. Interesting questions for further research are e.g. the mathematical implication by different factors between parameter classes, the reliability of the results, ideally in comparison to the results of quantitative risk assessment methods and the best presentation of the parameters and its classes.

References

1. Milius, B.: A New Classification for Risk Assessment Methods. In: Proceedings of 6th Symposium FORMS/FORMAT 2007 - "Formal Methods for Automation and Safety in Railway and Automotive Systems" 25th - 26th January 2007.
2. Functional safety of electrical/electronic/programmable electronic safety-related systems - Part 1: General requirements (IEC 61508-1:1998 + Corrigendum 1999); German version EN 61508-1:2001 /bibitem14.3 Verband Deutscher Verkehrsunternehmen: VDV 332: Sicherheitsintegritätsanforderungen für Bahnsignalanlagen bei Nichtbundeseigenen Eisenbahnen (NE). Juli 2008
3. Milius, B.: Konstruktion eines semi-qualitativen Risikographen für das Eisenbahnwesen. Dissertation. December 2009. `http://rzbl04.biblio.etc.tu-bs.de:8080/docportal/receive/DocPortal_document_00031882`
4. Railway applications - The specification and demonstration of reliability, availability, maintainability and safety (RAMS); German version EN 50126:1999
5. Braband, J.: Risikoanalysen in der Eisenbahn-Autommatisierung. Herausgegeben von der Siemens AG, 2005
6. Milius, Birgit: Severity of Harm in Semi-Quantitative Risk Assessment Method. In: Proceedings of the Fourth International Conference on System Safety 26 -28 October 2009, Savoy Place, London, UK.
7. Union Internationale des Chemines de Fer: Generic Hazard List Methodology for Railway Signaling, 2007.
8. Commission regulation (EC) No 352/2009 of 24 April 2009 on the adoption of a common safety method on risk evaluation and assessment as referred to in Article 6(3)(a) of Directive 2004/49/EC of the European Parliament and of the Council

On the Justification of a Risk Matrix for Technical Systems in European Railways

Jens Braband

Siemens AG, Industry Sector, Mobility Division, Rail Automation
Ackerstr. 22, 38126 Braunschweig, Germany
jens.braband@siemens.com

Abstract. The European Railway Agency (ERA) has the challenging task of establishing common safety targets (CSTs) and common safety methods (CSMs) throughout Europe. In this context, the harmonization of risk matrices is also discussed. The purpose of this paper is to provide a formal justification of risk matrices for technical systems and the means by which compliance with legal and regulatory requirements can be demonstrated. A proposal for a standard risk matrix applicable to technical systems is derived.

Keywords: Common Safety Method (CSM), Risk Analysis, Risk Matrix, CSM Regulation, European Railway Agency (ERA).

1 Introduction

The European Railway Agency (http://www.era.europa.eu), established by European Regulation 881/2004, has the mission of reinforcing railway safety and interoperability throughout Europe in times of ongoing privatisation. Central to its work on railway safety is the development of measures based on common safety targets (CSTs) and common safety methods (CSMs), common safety indicators and harmonised safety certification documents.

The common safety methods describe how safety levels, the achievement of safety targets and compliance with other safety requirements are assessed in the various Member States. As a first step, EU Regulation 352/2009 came into force in July 2010.

In this regulation, a semi-quantitative risk acceptance criterion for technical systems (RAC-TS) has been included: *For technical systems where a functional failure has credible direct potential for a catastrophic consequence, the associated risk does not have to be reduced further if the rate of that failure is less than or equal to* 10^{-9} *per operating hour.*

This criterion is limited to those technical systems where the failure can lead to catastrophic effects, e.g. accidents involving several fatalities, and for which there are no credible barriers or substantial mitigating factors that will prevent this consequence from materialising. The criterion can be used for the most critical functions performed by technical systems on railways such as

E. Schnieder, G. Tarnai (eds.), *FORMS/FORMAT 2010*,
DOI 10.1007/978-3-642-14261-1_19, © Springer-Verlag Berlin Heidelberg 2011

speed supervision, control of point position, complete and permanent loss of the brake system, or loss of the traction cut-off function. Further explanations are given by Braband (2007).

This means that formally RAC-TS relates only to potentially catastrophic accidents. In order to apply it also to other severity categories, for example, a risk matrix could be used. However, such a matrix would need sound justification.

This paper seeks to give formal justification in order to move a step forward towards the harmonisation of risk matrices.

2 Risk Matrix

Risk matrices are a well-known tool in risk assessment and risk classification, also in the railway domain (see for example EN 50126 (1999) or Braband (2005)). Table 1 gives a typical example. The major drawbacks of such risk matrices are:

- Risk matrices must be calibrated to their particular application.
- The results depend on the system level to which they are applied.
- The parameter classes must be concisely defined in order to avoid ambiguity and misjudgements.
- It must be defined which frequency is meant, e.g. accident or hazard frequency.

However, if these drawbacks can be overcome, risk matrixes are a well-accepted and easy-to-use tool, also for risk prioritisation (see the rank numbers in Table 1).

Table 1. Typical Risk Matrix

Frequency				
Often	10	6	3	1
Probable	14	9	5	2
Occasional	18	13	8	4
Rare	21	17	12	7
Improbable	23	20	16	11
Unbelievable	24	22	19	15
	Negligible	Marginal	Critical	Catastrophic
	Severity			

3 System Definition

The discussion in this paper focuses on technical systems only. A technical system is a product developed by a supplier including its design, implementation, and support documentation. It should be noted that:

- The development of a technical system starts with its system requirements specification and ends with its safety approval.
- Human operators and their actions are not included in a technical system.
- Maintenance is not included in the definition, but maintenance manuals.
- Technical systems can be subject to a generic type approval, for which a stand-alone risk acceptance criterion is useful.

A function is a "specific purpose or objective to be accomplished that can be specified or described without reference to the physical means of achieving it." A function level is defined in prEN 15380-4 (2010) as "level, to group functions of equal purpose". The distinction between levels is described informally as follows:

- First-level function: functional domain that encompasses a set of functions related to the same general focus or service for the considered (rolling stock) system.
- Second-level function: related to a specific set of activities which contributes to completion of the functional domain defined at the first level (at this level, it is not said how a second-level function is to be implemented).
- Third-level function: related to a specific activity out of the related set of activities, it encompasses a set of tasks (a function at least at level 3 should be supported as much as possible by one single subsystem).

It is proposed to use prEN 15380-4 (2010) which contains up to five hierarchical levels. Taking into account the definition of function level, level 3 seems to be the most appropriate for the application of RAC-TS. At least it does not seem reasonable to go into more detailed levels such as level 4 or 5. Table 2 gives some examples of functions to which RAC-TS may be applied. Although prEN 15380-4 (2010) relates to rolling stock only, it can be extended to infrastructure functions quite easily, e.g. by identification of all interfaces of other functions to rolling stock. Some functions (or at least interfaces) are already included in group K. In Table 2, some examples of level 3 functions related to signalling are proposed.

4 Severity Classification

The first step is to find a proper approach towards the classification of accident severities. Here, EC Regulation 352/2009 has already made the first decision and has defined the severity in RAC-TS qualitatively. This seems reasonable as the severity of an accident scenario contains a strong random element. Two very similar scenarios, e.g. the derailment of a passenger train at high speed, may in reality lead to very dissimilar outcomes: while one accident may result in dozens of fatalities, the other may result only in a few light injuries. Also, due to the high safety level of European railways, such accidents occur very rarely, so that the basis for statistical estimation

is questionable. However, numerical values may be used for the validation of severity categories.
It is natural to classify railway accidents by the following guiding questions:

1. Does the accident scenario typically involve a single person (P) or multiple persons (M)?
2. Does the accident typically lead to light (L) or serious injury (S) or fatality (F) of a person involved?

Table 2. Examples of signalling functions

Code	Function description
=LBB	Detect track vacancy
=LBC	Detect train at a particular spot
=LBD	Locate train
=LCB	Determine train description
=LDB	Provide diagnostics
=LEB	Supervise driver vigilance
=LEC	Automatic train stop
=LED	Supervise braking curve
=LEE	Supervise maximum train speed
=LFB	Optimise train running
=LGB	Monitor points
=LGC	Lock points
=LGD	Monitor derailer
=LGE	Lock derailer
=LGF	Monitor level crossing
=LHB	Provide signal information
=LJB	Provide cab radio
=LKB	Display state to driver
=LKC	Display state to dispatcher
=LKD	Transmit commands

It should be noted that

- "Typically" does not mean worst case; in a safety meaning, it should be interpreted as a "typical" bad outcome, i.e. worse than average.
- If statistical data were available, e.g. a 90% percentile of the outcome distribution could be taken. This means that 90% of accidents have a less severe outcome, but 10% have a more severe outcome.
- f no trustworthy data or expert estimates are available, then a conservative choice has to be made, e.g. if an accident may typically lead to severe injury or fatality, but if no clear distinction is possible, then the category fatality has to be chosen

- Train accidents usually belong to category M while accidents sustained by persons belong to category P.

This already results in an initial classification of severity categories: M-F, P-F, M-S, P-S, M-L and P-L. In order to judge this classification, it can be compared to a common numerical scale, the so-called fatalities-and-weighted-injuries (FWI) scale, where it is often assumed that one fatality is equivalent to 10 serious injuries and 100 light injuries, respectively. Table 3 shows that this approach also seems numerically plausible and quite natural.

Table 3. Plausibility check for severity categories

Combination	FWI range	Typical	FWI EN 50126
M-F	2 FWI	5	Catastrophic
P-F	1 FWI<2	1	Catastrophic/ critical
M-S	0.2 FWI<1	0.5	Critical
P-S	0.1 FWI<0.2	0.1	Critical
M-L	0.02 FWI<0.1	0.05	Marginal
P-L	0.01 FWI<0.02	0.01	Marginal
-	FWI<0.01	n. a.	Insignificant

Often, proposals are made to add an additional severity class on top of combination M-F or to subdivide M-F into two subclasses (e.g. FWI<10 and FWI 10). However, this does not seem to be practicable for setting design targets for technical systems for the following reasons:

- The outcome in terms of FWI for an accident of combination M-F is more or less random, so it would be very difficult to distinguish between further subclasses
- EC Regulation 352/2009 already sets the most demanding safety target for a potentially catastrophic scenario. So, for the process of setting design target, a further distinction is not needed.

5 Assessment of Accident Scenarios

In the next step, it must be checked whether the classification of accident scenarios with respect to the proposed severity categories is feasible. In order to classify accident scenarios, two different approaches are feasible:

1. Expert judgement has to be treated with care as experts often seem to remember worst cases well and tend to overestimate typical bad outcomes
2. The statistical evaluation of accident data often has problems in that the data for severe accidents is rare and that sometimes data quality is questionable

Thus, it seems good advice to use both approaches (where possible) and to use one method as a plausibility check for the results of the other method. The optimal approach would have to be a fixed assignment from accident scenarios to severity classes so that the variations in expert judgements are minimised. However, already from Table 2, it is apparent that classes P-F and P-S are quite narrow and may be hard to be distinguished from their neighbours. From this observation, the consolidated severity categories shown in Table 4 result.

Table 4. Consolidated severity categories

ID	Combinations	FWI range	Typical FWI
E	M-F	2 FWI	5
D	P-F, M-S	0.2 FWI<2	1
C	P-S, M-L	0.02 FWI<0.2	0.1
B	P-L	0.01 FWI<0.02	0.01
A	-	FWI<0.01	n. a.

In Table 4, simple letters are used to denote the severity categories and unambiguous, sometimes also misleading, verbal descriptions.

Table 5 shows an example for a simple classification of accidents with respect to their severity. With such a table, an unambiguous and easy-to-use classification would be possible. Similar classification tables are already in use in the car industry, e.g. the new draft ISO 26262 safety standard (2010).

Table 5. Examples of accident classification

ID	Derailment...	Collision...	Impact	Personal accidents
E	Passenger train at high speed	Passenger train on a main line		-
D	Passenger train at medium speed	At a level crossing	Train with work gang	Passenger falling out of a train at high speed
C	Passenger train at low speed			Passenger falling out of a train at low speed or at stop
B	In shunting operation		Train into buffer at low speed	Passenger hit by a door Passenger falling during embarkment
A				

6 Assignment of Frequencies to Accident Categories

In order to derive design targets for technical systems where functional failure has direct consequences, mapping from the class ID to the tolerable frequencies of the scenario is necessary. The generalisation from RAC-TS, which corresponds to class ID E, to other severity classes, can be done by extrapolation or by examples of design criteria for technical systems. In this context, also effects such as risk aversion have to be taken into account.

In such a mapping, two fixed points exist: the tolerable hazard rate of 10^{-9} per operating hour per function of a technical system (from RAC-TS) relates to class ID E and the assignment of a hazard rate of less than 10^{-5} to class ID A, as it is generally accepted that, if an accident scenario does not lead to injuries also no safety requirements should then be demanded from the function of the technical system. Thus, an even decadal assignment of rates to severity classes is not possible as the frequency bandwidth is at least one order of magnitude too wide.

Now, risk aversion can be taken into account which means that more severe safety requirements should be imposed on functions that may result in more severe consequences. This means in this context that the spread of the frequency bandwidth may be wider for lower severity and narrower for higher severity. Thus, it is proposed to use a decadal proportion for the upper severity classes and a more relaxed proportion for the lower severity classes (see Table 6).

Table 6. Proposed risk matrix

HR	A	B	C	D	E
$> 10^{-5}/h$					
$10^{-5}/h$				intolerable	
$3 \cdot 10^{-7}/h$					
$10^{-8}/h$	tolerable				
$10^{-9}/h$					RAC-TS

In practice, for the allocation of HR to hazards, only the diagonal of the risk matrix is necessary, so it may also be represented as a risk table (see Table 7). As a further plausibility check, the corresponding SIL is also stated. However, it should be noted that this SIL would only apply if there were no credible risk reduction factors or barriers at all, so that the accident would directly occur whenever the hazard occurs.

7 Examples

In some cases, like =LGB from Table 2, RAC-TS is directly applicable. The main hazard would be that the status of points would be determined wrongly

Table 7. Proposed risk table

ID	HR	SIL equivalent
E	$10^{-9}/h$	4
D	$10^{-8}/h$	3
C	$3 \cdot 10^{-7}/h$	2
B	$10^{-5}/h$	1
A	$> 10^{-5}/h$	n. a.

so that a train may run over points which are set in an incorrect direction. If passenger trains at high speed ran over these points, then ID E would be determined from Table 5 leading to a THR of 10^{-9} per operating hour per set of points.
In another example, =LGF from Table 2, the main hazard would be that road traffic would not be protected by the level crossing and the consequence might be a collision at the level crossing, from which ID D would derive from Table 5 leading to a THR of 10^{-8} per operating hour per level crossing. Note that usually these THRs are reduced by risk reduction factors or mitigation barriers, so that normally the THRs derived from the risk matrix are not the THRs for the technical functions.
For accident types not classified in Table 5, either classification could be done by analogy or statistical data or expert judgement could be applied directly. For example, if it were judged that a particular accident typically results in a single fatality, then ID D would be chosen based on Table 4. If statistical evaluation resulted in a FWI value of 0.05 with 90% confidence, then the ID C should be chosen from Table 4.

8 Conclusions and Outlook

The risk acceptance and setting of THR for technical systems could be based on a risk matrix as explained in this document. In a next step, it is possible to take a set of typical barriers into account. Some examples of how to fulfil RAC-TS with the risk matrix are given in this document.
When using the risk matrix, mutual recognition will also depend on the list of functions to which the risk matrix is applied. So, the use of a common risk matrix will facilitate the mutual recognition process, but not lead to an automatic approval.

References

1. Braband, J. (2005). Risikoanalysen in der Eisenbahn-Automatisierung, Tetzlaff, 2005
2. Braband, J. (2007): A proposal for common safety methods for risk assessment in European Railways, Signal&Draht, 4/2007, 34-37

3. CENELEC (1997) EN 50126 Railway applications -The specification and demonstration of Reliability, Availability, Maintainability and Safety (RAMS)
4. CENELEC (2010) prEN 15380 Part 4: Railway applications - Classification system for rail vehicles - Function groups
5. EC (2009) Regulation No. 352/2009 of 24 April 2009 on the adoption of a common safety method on risk evaluation and assessment as referred to in Article 6(3)(a) of Directive 2004/49/EC of the European Parliament and of the Council
6. ISO (2010) DIS 26262: Road vehicles - Functional safety

Using Guided Simulation to Assess Driver Assistance Systems *

Martin Fränzle[1], Tayfun Gezgin[2], Hardi Hungar[2], Stefan Puch[1], and Gerald Sauter[1]

[1] Carl von Ossietzky Universität Oldenburg,
Department of Computer Science, 26111 Oldenburg, Germany,
{fraenzle,stefan.puch,gerald.sauter}
@informatik.uni-oldenburg.de
[2] OFFIS, Escherweg 2, 26121 Oldenburg, Germany,
{hungar,gezgin}@offis.de

Abstract. The goal of our approach is the model-based prediction of the effects of driver assistance systems. To achieve this we integrate models of a driver and a car within a simulation environment and face the problem of analysing the emergent effects of the resulting complex system with discrete, numeric and probabilistic components. In particular, it is difficult to assess the probability of rare events, though we are specifically interested in critical situations which will be infrequent for any reasonable system. For that purpose, we use a quantitative logic which enables us to specify criticality and other properties of simulation runs. An online evaluation of the logic permits us to define a procedure which guides the simulation towards critical situations and allows to estimate the risk connected with the introduction of the assistance system.

Keywords: Guided Simulation, Formal Specification, Model Integration, Model-based Design, Assistance System, Driver Modelling

1 Introduction

The design of an assistance system in the automotive domain (and elsewhere) requires several exploration and evaluation activities with potential users of the system to assess the effect of the system under development. As a consequence, the development process is difficult to organize, it is expensive and time consuming. The goal of the IMoST[3] project is to reduce the amount of involvement of human test subjects through the introduction of executable

* The research reported here has been mainly performed in the project IMoST which is funded by the Ministry of Science and Culture of Lower Saxonia.

[3] The full name of the project is "Integrated Modelling for Safe Transportation". Further information can be found in [1] and at the URL http://imost.informatik.uni-oldenburg.de/

E. Schnieder, G. Tarnai (eds.), *FORMS/FORMAT 2010*,
DOI 10.1007/978-3-642-14261-1_20, © Springer-Verlag Berlin Heidelberg 2011

models of the driver. These models shall be able to replace the driver in that they are capable of reproducing human behaviour. Combining them with executable models of the car, traffic scenario, and the assistance system, a complete operational representation of the assistance system in its application environment can then be constructed and employed to predict effects of introducing the assistance without having to resort to experiments with humans. While the construction of driver models is a both scientifically and practically challenging task which is addressed in a number of other reports, e.g. [5,6,8], in this paper we focus on techniques concerned with using these models, i.e., with evaluating functionality and safety aspects of driving with assistance. The evaluation is performed by studying the emergent behaviour of the integrated models. As the models are rather complex, the main means for assessing them must be simulation, because other analysis methods (e.g., computing all states the model may reach or even formal verification) are only applicable to much simpler classes of systems or smaller models.

Of course, the simulation activity must be well organized to produce reliable assessments. Our approach combines a systematic parameter coverage with property-specific guidance. If, for instance, we are interested in a particular aspect of criticality, we start with a function assigning a numeric criticality value to each run. After covering the parameter space roughly, areas where high values have been observed get analysed in more detail. Thus, the simulation proceeds, guided by observations, towards points of interest. In particular, the (hopefully low) probability of situations like those involving a high accident risk can be assessed with much greater accuracy than by simpler procedures.

The application scenario on which IMoST develops and tests its approach is that of an advanced driver assistance system (ADAS) supporting the driver in filtering into an expressway, including the gap selection and speed adaptation. This scenario captures one of the most critical expressway manoeuvres. On the other hand, compared to other potentially critical traffic situations (e.g., crossings), it is limited in its variability and is thus suited for developing and assessing a new approach. Variables we considered were the number of other traffic participants, speed differences, and gap sizes.

This paper is organized as follows. In the following section we present our simulation platform, i.e. list the components of the considered co-simulation and state how the interaction between all components takes place. In Sec. 3 we give a formalism allowing to specify (among other) safety properties. The succeeding section deals with the online evaluation of the specified formulas. In Sec. 5 we will give a procedure to automatically determine critical situations, and we conclude with Sec. 6.

2 Simulation Platform

The complete model consists of several software modules. These are provided from different sources: They incorporate a commercial traffic simulation, the models of the driver and the assistance system developed in the project and components for monitoring and recording. The most convenient way to cope with continuing changes of these modules is to refrain from a deep integration into one system and rather combine them via a co-simulation environment. For that purpose, we use a commercial implementation of the IEEE standard 1516 [3] for coupling simulators (HLA, "High Level Architecture"). This standard defines how a joint run of different component simulators is orchestrated by a central component (RTI, "Run-Time Infrastructure").

The HLA term for a set of combined simulators is *federation*, and each partner is called a *federate*. HLA offers a time management service which enables to synchronize federates running at different and even variable step resolutions. A federate is *time regulating* if it influences the advance of other federates, and it is *time constrained* if its own evolution is restricted by others. Time management permits to keep the data exchange in accordance with the progress of logical time, opposed to best-effort simulation where data are consumed as they become available during simulation. To limit variation between different simulation runs with the same parameters, i.e., to achieve a high degree of reproducibility, we used this time management. For technical reasons, in particular the nature of the commercial traffic simulation software, even this does not suffice for full reproducibility. It is, though, planned to replace that component with another one what we expect to remove these problems.

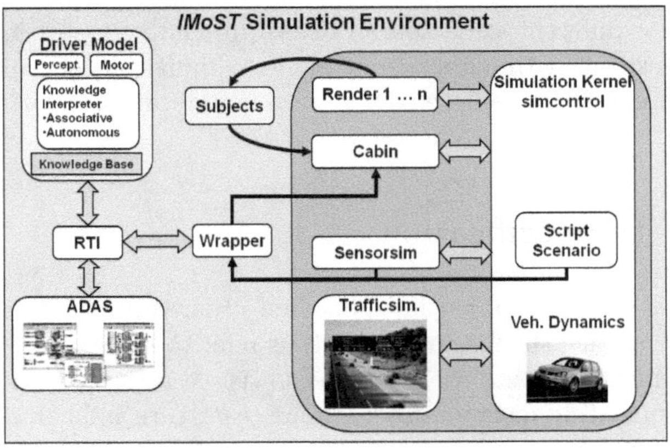

Fig. 1. Architecture of the federated simulation.

Fig. 2 depicts the structure of the main components and their integration by the RTI. In particular, the main components are models of the driver and assistance-system (Advanced Driver Assistance System, ADAS) on the left of the figure and a simulation of the ego car, which is the car controlled by the driver model, and the traffic environment on the right. Further components not shown in the figure are property monitors (or observer, see Sec. 3 for details) and a recorder. The co-simulation is executed on a cluster of standard PCs.

Each component model evolves in discrete time steps, with update frequency resolutions in the order of 20 to 35 Hz. Synchronisation and data exchange is managed by the RTI. The output behaviour of models of the ADAS, the ego car and the environment depends deterministically on their input, where the traffic environment is parameterized by scripts defining the street layout and number and actions of other cars. A complete run of the scenario consists on average of about 2700 discrete time steps.

The IMoST cognitive driver model consist of two parts: 1) The cognitive architecture CASCaS, which integrates task-independent human cognitive processes, e.g. a model of visual perception, declarative and procedural memory models and a processor for task knowledge. 2) A formal model of driving-task specific knowledge (e.g. knowledge about different driving manoeuvres or traffic rules), which is "uploaded" onto the architecture for simulation purposes. Thus, a cognitive architecture can be understood as a generic interpreter that executes task-specific knowledge in a psychologically plausible way.

The driver model incorporates different types of behavioural variation concerning acceleration style (sportive vs. relaxed), gaze strategies (variations in gaze duration and frequency) and safety margins (preferred distance to lead car / rear car), which where assessed through a series of experiments with subjects. During simulation the model probabilistically chooses amongst those different behaviours. These probabilistic capabilities are of specific interest when thinking about guided simulation, because the probabilistic choice will be replaced by a systematic variation of the possible behaviours.

3 Property Specification

The properties of interest are defined in a first-order version of linear temporal logic. In their atoms, the formulas may thus refer to attributes of the system constituents like car positions, their speed, visible actions of the driver and so on, complementing discrete observations (e.g., turn indicator, assistance-system signals). The usual temporal operators (always, eventually, unless, until) permit to express temporal relations of their occurrence. Formulas are evaluated over complete traces which usually originate from the simulation environment but might also be recordings of driver tests. With the temporal operators one can formulate specific requirements on different situations and phases of driving.

Extending the usual interpretation of logics, we chose a nonstandard, quantitative semantics [2], [7] which assigns a numerical value to a formula for each trace: A positive number means that the formula is satisfied, and the result value gives the minimal distance in variable values which would have made the formula false (conversely for negative values). Thus, a formula defines a function which assigns a numerical value to each simulation run. Due to this nonstandard interpretation, numerical assessment functions can be expressed in the logic.

As an example, we use formulas to describe (aspects of) criticality of simulator runs. With an adequate subformula defining "TimeToCollision", computed from distance and relative speed of a leading car, if there is any, the formula

$$\square(\mathit{TimeToCollision} > 2.6sec) \tag{1}$$

expresses that this value never drops below 2.6 seconds (which would be rather safe). Usually, we want more than one aspect evaluated. As an example, also "TimeHeadway" enters criticality, and an acceptable lower bound on this value in our scenario is 0.3 seconds. To provide an adequate criticality criterion in one formula, both observations are transformed into a *risk* valuation. The risk is inversely proportional to the respective time values, and with adequate scaling factors we arrive at the following definitions.

$$\mathit{RiskTTC} =_{\mathrm{df}} 2.6/\max(\mathit{TimeToCollision}, 0.01) \tag{2}$$

$$\mathit{RiskTHw} =_{\mathrm{df}} 0.3/\max(\mathit{TimeHeadway}, 0.01) \tag{3}$$

These formulas yield 1 for the values of 2.6, resp., 0.3, seconds, and higher values (with an upper bound of 100) for tighter situations. A criticality criterion for our scenario is specified by

$$\mathit{OnAccelLane}\,\mathsf{W}\,(\neg \mathit{OnAccelLane} \wedge \mathit{RiskTTC} \leq 1 \wedge \mathit{RiskThw} \leq 1)\ , \tag{4}$$

with W for "unless". If this formula evaluates to a positive value for a run, the driver has performed the manoeuvre with sufficient safety margins in every respect. A value of -2 indicates an already severe criticality with only 4.8 meters distance at a velocity difference of 20 km/h (*RiskTTC*) or 0.1 seconds headway – only 2.2 meters at 80 km/h (*RiskTHw*).

The evaluation of formulas is automated by translating them into monitor programs [4] observing to which degree the property is satisfied or violated. These monitors enter the simulation environment as additional federates. Upon termination of the simulation, each observer provides the numerical evaluation of the property it stands for on the completed run. Thus, the run can be classified as good or bad according to the resulting numbers. Moreover, the observers are capable of computing lower and upper bounds for the final value while the system is evolving. This allows us to identify and stop irrelevant runs at an early stage of evolvment in order to save simulation resources. The evaluation process is presented in more detail in the following section.

4 Online Evaluation of LTL formulas

Our goal is to evaluate (linear) temporal-logic formulas during simulation, i.e. over traces while they are evolving. The evaluation has to be performed in

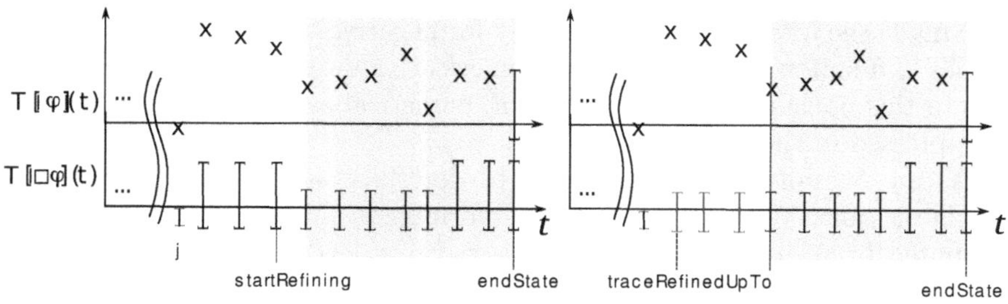

Fig. 2. Refinement of the *always* operator. The shaded area is the last evaluated frame. Left: Evaluation before refinement; Right: Evaluation after refinement.

real-time in order to provide the result timely. As mentioned in the previous section we evaluate such formulas quantitatively. We face two main problems in this evaluation process: First, the nature of the temporal operators requires that in order to evaluate the truth value (or degree) at one time instant we need the truth values of all future time instants. The second one is that the semantics of such formulas are defined over infinite traces, so we have to find a consistent interpretation on *finite* traces.

The first problem is solved in the manner that traces are safely extrapolated for all future time instants. This is realized by interpreting attributes over intervals rather than single values. For all time instants where a measured value for an attribute is already available we get a singleton interval. If we have no further information about the evolution of an attribute, we have to extrapolate its value for all time instants without measurements with $[-\infty, \infty]$. By annotating an attribute with a monotonicity information, a tighter extrapolation can be performed: If an attribute is monotonically increasing the last measured value is used as the lower bound for all future time instants. For monotonically decreasing attributes the upper bound is set analogously. Whenever new measurements are available, the values of each previously evaluated time instant are refined.

The interval based evaluation also solves the above mentioned second problem. At the end of each simulation run, the given formula f is evaluated by an interval $[lb, ub]$ with $lb, ub \in \mathbb{N}$ which is interpreted as follows: If $lb > 0$, then all possible extensions of the observed trace will satisfy f. Analogously, if $ub < 0$ then all extensions of the trace will not satisfy f. The evaluation may result in an indefinite answer if the interval includes zero.

For our implementation it is crucial that the extrapolation for all future time instants can be finitely represented. In the following, we sketch the evaluation procedure.

The evaluation of a formula is realized bottom-up (wrt. the formula structure) and in a depth-first search manner (wrt. the states to be checked). The online checking procedure is performed iteratively due to the evolving nature of the trace. During a checking iteration the trace is not changed, i.e. newly measured values are put into a buffer. Before the next iteration starts, the trace is extended accordingly. A checking iteration consists of two steps:

- *Evaluation Phase*: In the *evaluation phase* all subformulas and all states from a start state up to an *end state* are sequentially computed. The end state is determined by the largest time instant for which a value for all attributes exists. For states beyond this last state the extrapolated interval, as explained above, is used. The evaluated area will be called *frame*. The start state for the next checking iteration is the last checked state.
- *Refinement Phase:* The evaluation of the last frame results in new information of attribute behaviours. States already evaluated in previous iterations can thus be refined. This is done by backward iterating all previously evaluated states until no further refinement on values can be performed.

The evaluation of temporal operators is realized according to their recursive characterizations. In the following, the idea of the procedure is shown for the *always* operator. Let T be an (extrapolated) trace, v be an attribute and $i \in \mathbb{N}$ a time instant. The evaluation of v at time i then is denoted by

$$\forall i \in \mathbb{N}: \; T[\![\Box\varphi]\!](i) = T[\![\varphi]\!](i) \wedge T[\![\Box\varphi]\!](i+1). \tag{5}$$

In order to be able to compute the formula values correctly, we store the current evaluation results for all time instants of all attributes and subformulas. That this information is needed becomes clear when one considers how to evaluate formulas with nested temporal operators like $\Diamond(a > c_1)\ U\ (b > c_2)$. The refinement phase concerns the update of these preliminary values, reflecting the effect of tightening intervals extrapolated previously. Note that the final result is the evaluation of the topmost formula operator in the first state.

The idea of the refinement process is shown in the example of Fig. 2: In the figure, the truth values starting from state j up to state `endState` of both formulas, $f_1 = \varphi$ and $f_2 = \Box\varphi$ are depicted. The left part of the figure illustrates the result after the evaluation phase for the shaded area. The effect of the subsequent refinement phase is demonstrated in the right part of the figure, leading to tighter intervals up to state `traceRefinedUpTo`. The information about changed values of subformulas during refinement is of course important for the refinement of the enclosing formula.

An exact evaluation of formulas according to the procedure sketched above is rather costly. In order to reduce the state space, one can merge multiple measured values into one if they are very similar, e.g., their values only differ by a fixed threshold. In practice, this has been proven to be very useful, as measurements arrive with a very high frequency while the corresponding values do only differ slightly. To illustrate the effects of such merging techniques, we have specified and evaluated different formulas which refer to the attributes of the car in our application scenario. The results are illustrated

Formula	(i)	(ii)	(iii)	Evaluation Result
$f_1 = \Box\Diamond v > 0$	2600	1200	630	$[26,\ D_{max}]$
$f_2 = \Diamond(d > 0 \wedge lane = right)$	2710	1190	710	$[0.1,\ D_{max}]$
$f_3 = v < 30\ U\ d > 0$	2700	1190	710	$[5.49,\ 5.49]$
$f_4 = v < 30\ U\ f_2$	3000	1640	800	$[0.104,\ D_{max}]$

Table 1. Online evaluation of formulas in the context of the IMoST scenario. Here d is the distance to the acceleration lane and v the speed of the considered car (in $\frac{m}{s}$). Both d and v are monotonically increasing. Size of state spaces resulting from (i) no merging, (ii) value threshold and (iii) time threshold. D_{max} is the representation of ∞.

in Table 1. The effects on the state spaces using different merging techniques are shown in the central columns: In Column (i), no merging was applied at all, such that the state spaces grew very fast. E.g., to evaluate formula f_1, 2600 states were processed. In (ii), states where attribute values differed by less than 0.1 were merged. Finally in (iii), the merging was performed with respect to the time stamps of the measurements, i.e., measurements with differences in their time stamps less than $10ms$ were merged. As a result, by using these merging techniques, far less states had to be treated by our evaluation procedure, with of course direct effects on the computation time needed for the refinement procedure.

The evaluation results of the formulas are listed in the rightmost column of the table. Merging affected the results only negligibly, wherefore we did not list these effects. We will consider how to capitalize in practice on the substantial optimization potential of these techniques.

5 Guided Simulation by Exploring the Behaviour Spectrum

It should be obvious that the variability of the scenario is too high to cover all its instances by simulation, let alone by experiments with human subjects. Also, the probabilistic nature of the driver model does complicate matters. To overcome this problem, we propose the following approach:

- A property of interest is specified by a temporal-logic formula.
- A batch of simulations is performed with the intent of roughly exploring the spectrum. For that, test points covering the value ranges of the scenario parameters are chosen. This batch of simulations provides a grid of sampling points for the property function from which an approximation of the function is derived by interpolation.
- At each test point, the probabilistic property function is evaluated either by Monte-Carlo simulation or, more elaborately, by systematically exploring the behaviour spectrum of the driver model (see below for details).

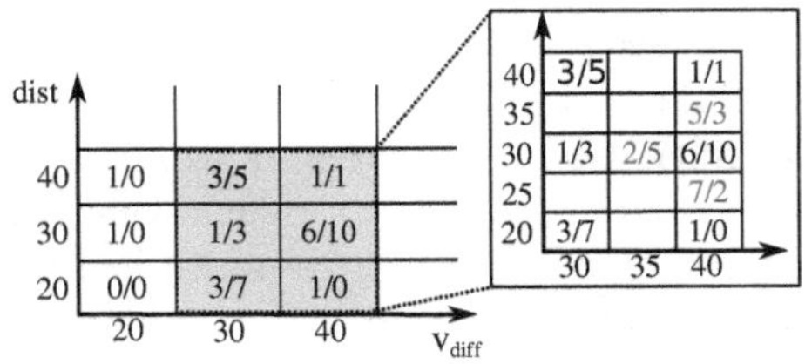

Fig. 3. Evaluating criticality at test points in a scenario with parameters v_{diff} (speed difference to cars on the right lane, [km/h]) and *dist* (gap size, [m]).

– Further simulation refines the approximation in areas of interest (e.g., in regions with high function values) by selecting input values accordingly.

Via such a guided simulation, maxima (or minima) of the property function can be detected with far less simulation runs than by brute force. We have instantiated this approach for our application scenario, where indeed the main factors determining the course a simulation run takes are the parameters of the traffic scenario to be explored and the decisions the driver model takes in reaction to the scenario. We will not describe here the adaptations necessary to cope with the nondeterminism occurring in practice, which results from, e.g., race conditions.

Results of an exploration where we applied Monte-Carlo simulation at each test point are depicted in Fig. 3. The property used is defined by the formula (4) from Sec. 3. It yields negative values for critical runs, so we seek minima of this function. We considered variations of the attributes $v_{diff}\ [km/h], dist\ [m] \in \{20, 30, 40\}$, where *dist* denotes the size of the gap size on the right lane of the expressway to be used by the ego car, and v_{diff} is the speed difference to the cars on the right lane at the point in time when the ego car enters the acceleration lane. Each combination of values of these attributes yields a test point, and 20 simulation runs were performed for each point. Entries in the result matrix are the numbers of *unacceptable* and *severely critical* runs, defined by formula values below -10 or in [-10,-2], respectively. For instance, we get 6 unacceptable and 10 severely critical runs for the test point $v_{diff} = 40\ km/h$ and $dist = 30\ m$. These rather high criticalities have their reason in the fact that, with these parameters, the scenario is very demanding and that the driver model we used was not yet fully adapted to the scenario. The right part of Fig. 3 shows how the criticality function is refined in the vicinity of the point with highest observed criticality.

A better evaluation of the property value at a test point can be achieved by replacing the Monte-Carlo simulation by a systematic exploration of the probability space. This is possible in our setup as we can control the probabilistic decisions of the driver model externally. Depending on the history of the simulation, the driver model reaches points where it uses random numbers to chose between different courses of action. Conceptually, this yields a tree of possible runs for each set of scenario parameters. Each node in the tree stands for a probabilistic decision, and each edge is labelled accordingly by a probability. By its nature, this random tree is accessible (only) in a top-down fashion. To explore it, paths are taken systematically. The branch probabilities encountered along the way are multiplied to compute the path probability. If a path probability gets below a minimum threshold fixed at the beginning of the procedure, its further exploration is stopped. Completed runs yield values for the property of interest. These are used to annotate the branches which have been taken with estimations of (maximal) property values and reliability information, guiding the further exploration of the tree.

An implementation of this exploration procedure is currently under development, but still in an experimental state. Due to intricacies of the driver model and the simulation environment, further stabilization is needed to get a procedure yielding highly valid and dependable estimations.

6 Summary

We have presented a way of exploring via simulation the functionality of assistance systems and their effect on safety, given executable behaviour models of the driver and all other constituents of the scenario. The presented criticality-guided simulation explores the complex model and provides the designer with meaningful information on potentially dangerous situations arising from the current ADAS design and thus increases the quality. Our results on the presented case study indicate that this approach may indeed be helpful to reduce the number of tests with human subjects. The techniques are yet to be explored on a larger scale, which we intend to do in the near future. Also we will develop and test further techniques for speeding up the simulation and guaranteeing a high reliability of the resulting assessment. In particular, we will use the information about the internal states of driver and ADAS model for coverage and guidance.

Acknowledgements: We acknowledge the many fruitful discussions and in particular the work of the other participants in the IMoST project and further cooperating projects which provided the models whose behaviour we have set out to explore.

References

1. M. Baumann, H. Colonius, H. Hungar, F. Köster, M. Langner, A. Lüdtke, C. Möbus, J. Peinke, S. Puch, C. Schiessl, R. Steenken, and L. Weber. Integrated modelling for safe transportation - driver modeling and driver experiments, In: Fahrermodellierung in Wissenschaft und Wirtschaft, 2. Berliner Fachtagung für Fahrzeugsmodellierung. VDI Verlag, Düesseldorf, 2008.
2. L. de Alfaro, M. Faella, and M. I. A. Stoelinga. Linear and branching metrics for quantitative transition systems. In *Proc. 31st Int'l Colloq. on Automata, Languages and Programming (ICALP04), Turku, Finland*, volume 3142 of *Lecture Notes in Computer Science*, pages 97–109, Berlin, 2004. Springer.

3. IEEE 1516-2000 Standard for Modeling and Simulation (M&S). High level architecture - framework and rules. IEEE Computer Society Press, September 2000.
4. T. Gezgin. Observerbasierte on-the-fly Auswertung von QLTL-Formeln innerhalb eines HLA-Simulationsverbundes. Master's thesis, Carl von Ossietzky University Oldenburg, 2009.
5. Claus Möbus and Mark Eilers. Further steps towards driver modeling according to the bayesian programming approach. In Vincent Duffy, editor, *Digital Human Modeling*, volume 5620 of *Lecture Notes in Computer Science*, pages 413–422. Springer Berlin / Heidelberg, 2009.
6. Claus Möbus, Mark Eilers, Hilke Garbe, and Malte Zilinski. Probabilistic and empirical grounded modeling of agents in (partial) cooperative traffic scenarios. In Vincent Duffy, editor, *Digital Human Modeling*, volume 5620 of *Lecture Notes in Computer Science*, pages 423–432. Springer Berlin / Heidelberg, 2009.
7. Gerald Sauter, Henning Dierks, Martin Fränzle, and Michael R. Hansen. Lightweight hybrid model checking facilitating online prediction of temporal properties. In *Proceedings of the 21st Nordic Workshop on Programming Theory, NWPT '09*, pages 20–22, Kgs. Lyngby, Denmark, 2009. Danmarks Tekniske Universitet.
8. L. Weber, M. Baumann, A. Lüdtke, and R. Steenken. Modellierung von Entscheidungen beim Einfädeln auf die Autobahn. To appear: Fortschritts-Berichte VDI: Der Mensch im Mittelpunkt technischer Systeme. 8. Berliner Werkstatt Mensch-Maschine-Systeme, 2009.

The DeSCAS Methodology and Lessons Learned on Applying Formal Reasoning to Safety Domain Knowledge

Jan Gačnik[1], Henning Jost[2], Frank Köster[1], and Martin Fränzle[2]

[1] German Aerospace Center, Lilienthalplatz 7,
38108 Braunschweig, Germany
{jan.gacnik, frank.koester}@dlr.de
[2] University of Oldenburg, Ammerländer Heerstr. 114-118,
26129 Oldenburg, Germany
{henning.jost, martin.fraenzle}@informatik.uni-oldenburg.de

Abstract. Functional safety has become an important aspect for engineering activities in the automotive domain due to the upcoming introduction of the safety standard ISO 26262. This paper proposes a methodology to guide the safety related requirements engineering process by means of OWL (Web Ontology Language) ontologies. These ontologies formalize necessary domain knowledge and serve as reference models to support semi-automated requirements discovery and to ease the certification process. Using OWL's logical base, knowledge inference is applied to reason about safety measures for ensuring compliance with the reference process (guidance). The proposed methodology has been implemented in a prototype toolchain and applied to a simple lane departure warning system as an example assistance and automation system. Lessons learned refer to conceptual (expressiveness) and technical (tooling efficiency) issues.

Keywords: Certification, ISO 26262, Domain Knowledge, Ontology, Process Framework, Assistance and Automation System, Semantic Reasoning

1 Introduction

Safety critical systems like assistance and automation systems (AAS) in the automotive domain demand a clearly defined proceeding during development, especially to support certification and qualification processes. In order to reduce the risk of a hazardous system failure, standards have been defined which propose a certain proceeding, requirements and associated methods and measures during development. One of these standards is the upcoming ISO 26262 for functional safety in the automotive domain [6]. Due to the informal representation of such standards in natural language text, there is an inherent

E. Schnieder, G. Tarnai (eds.), *FORMS/FORMAT 2010*,
DOI 10.1007/978-3-642-14261-1_21, © Springer-Verlag Berlin Heidelberg 2011

risk of misinterpreting the standard's content or following an incorrect or incomplete sequence of development activities. This threat is further increased by interdisciplinary issues in developing safety critical systems: in order to avoid misunderstandings between different disciplines involved in the development, a universally comprehensible representation of system and process requirements – on a common and generic basis – is essential. In the context of ISO 26262, process requirements in particular are important, as most of the requirements from the standard are process related. A formally sound specification of process requirements based on a well-defined terminology is crucial for supporting a precise presentation of requirements. Furthermore, through the formal base, process requirements and associated activities become computer-readable, and some analysis steps (e.g. checking consistency) can be automated. Especially in the context of certification, traceability between process and system artifacts is of high importance.

1.1 Design of Safety Critical Automotive Systems

The *Virtual Institute DeSCAS* (Design of Safety Critical Automotive Systems), which is funded by the Helmholtz Association, strives for defining a process framework and related methods to support interdisciplinary development of safety critical assistance and automation systems in the automotive domain. The process framework builds on generic as well as domain dependent concepts to *interweave* different interdisciplinary development activities. This includes the formalization of relevant standards (e.g. ISO 26262) in order to allow the automation of certain analysis methods (e.g. hazard analysis and risk assessment) and the derivation of requirements, which may vary due to variations in safety level classification.

1.2 Structure of the Paper

This paper will outline how the use of a standard can be improved by formalizing its structure. This comprises OWL (Web Ontology Language) based formalization of the risk analysis from ISO 26262, which is applied to an example application (lane departure warning system). Since safety requirements and associated methods as well as process phases have been modeled in OWL, the result of the risk analysis is furthermore used to infer concrete requirements for a system under development. These requirements and their dependencies are utilized to derive a safety related workflow. The different steps of analysis and formal reasoning have been integrated in a prototype toolchain, which especially operates on safety requirements from ISO 26262. This paper gives a summary of the DeSCAS process framework concerning implementation aspects and insights gained from the prototype implementation (proof of concept). This especially concerns the performance of formal reasoning on OWL ontologies.

2 Proposed Methodology

Taking into account the current design and properties of state-of-the-art development in the automotive domain (such as ISO 26262 and RESPONSE 3 *Code of Practice* [8]), DeSCAS has developed a methodology for the development of assistance and automation systems. In addition to interweaving interdisciplinary development activities of the automotive domain by reasoning about domain knowledge, a formal base is essential for the DeSCAS process framework. Most of the state-of-the-art standards lack a formal representation as they primarily consist of glossary-based, natural language text descriptions, informal checklists or questionnaires, complemented by some graphics and tables. As a consequence, inconsistencies concerning both the use of technical terms and the dependencies between process elements become visible when analyzing these standards in terms of a formal definition. The nomenclature is partially unclear or even ambiguous, as are some connections between process elements and requirements [3]. For that reason, the process model as part of the entire DeSCAS process framework builds upon a generic and formally defined process meta-model.

2.1 Interweaving Development Streams

The essential parts of the DeSCAS meta-model are constituted by the *development streams* which combine related design activities of an interdisciplinary system development to be synchronized via iterations within a V-Model. Concerning the development of AAS in the automotive domain, the DeSCAS process model defines the following three main development streams:

1. *Human factors*: Continuously involving human behavior and interaction with the system during system development (see RESPONSE 3 *Code of Practice*).
2. *Functional development and architecture*: The classical hardware and software development of AAS is highly affected by the other two streams.
3. *Safety measures*: Compliance with standards such as ISO 26262 is inevitable for ensuring the functional safety of safety critical systems.

Interweaving these three development streams represents the interdisciplinary collaboration during the design of AAS. A detailed description of the DeSCAS process model and its process meta-model can be found in [4]. The knowledge deduction within the streams *functional development and architecture* and *safety measures* will be focused in the following by means of an example application.

2.2 Formalization of Domain Knowledge

Each development stream of the DeSCAS process model comprises stream-specific ontologies which precisely capture knowledge and expertise within a stream. In DeSCAS, OWL ontologies are used for the formal representation of such domain knowledge. OWL (Web Ontology Language) is a widely-used and open standard for describing ontologies and has been specified by the W3C[3]. The benefit and advantage of using OWL is its formal semantics which supports the application of a so-called *reasoner*. A reasoner is not only able to verify the consistency of the concepts modeled within an OWL ontology but also to perform logical deduction in order to automate analysis methods.

Domain standards such as ISO 26262 and the RESPONSE 3 *Code of Practice* supply domain terminology as in a glossary, reference process models and meta-models, as well as process- and product-related requirements. In order to show compliance of the development process to standards relevant for the system to be developed, these standards shall be formalized as done in DeSCAS with the committee draft of ISO 26262. The formalization of each standard involves modeling on the meta-model level and the model level:

- *Derive an ontology from the standard (meta-model level):* This ontology model is often related to the document structure of the standard. Concerning the committee draft of ISO 26262, about 60 meta-model concepts had to be modeled.
- *Model actual requirements and relations (model level):* This does not necessarily mean that the requirements need to be presented in a formal way (e.g. using formal languages) but rather documenting the relations and connections between several requirements and the relevant methods and measures, respectively, in a formalized way. The ontology of the committee draft of ISO 26262 exhibits up to 1000 instances of model concepts.

2.3 Semantic Reasoning

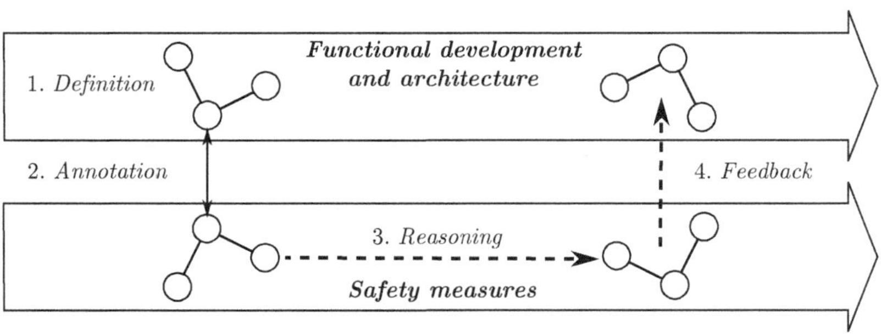

Fig. 1. Interweaving and reasoning within the DeSCAS methodology

Interweaving of models on the basis of formalized domain knowledge can be structured as follows, with Fig. 1 locating the steps within the respective development streams [2]:

[3] OWL – http://www.w3.org/TR/owl-ref/

1. *Definition* of a concrete domain specific model, which could be a functional requirements model for a certain system under development.
2. *Annotation* (direct or indirect) of the concrete model, based on formal domain ontologies: Thus, it is assured that requirements have a meaning in the sense of the respective domain ontologies.
3. *Reasoning* about implications within and from other domains: This could be an impact on safety and vice versa, due to a given definition of a system under development and associated safety measures.
4. *Feedback* of consequences from another development stream for the very own development activities: These could be concrete safety measures and related activities which either impact the requirements model or concurrent implementation activities.

3 Example Application and Prototype Toolchain

To further illustrate the proposed methodology, the development of an exemplified assistance and automation system from the automotive domain will be sketched as depicted in Fig. 2 and Fig. 3. Both the figure and the following textual description refine the four steps described in Sect. 2.3 – the enumeration of the subsequent text references the numbers in Fig. 3. This example application refers to a lane departure warning system (LDWS) which alerts the driver as soon as the vehicle begins to move out of its lane [5].

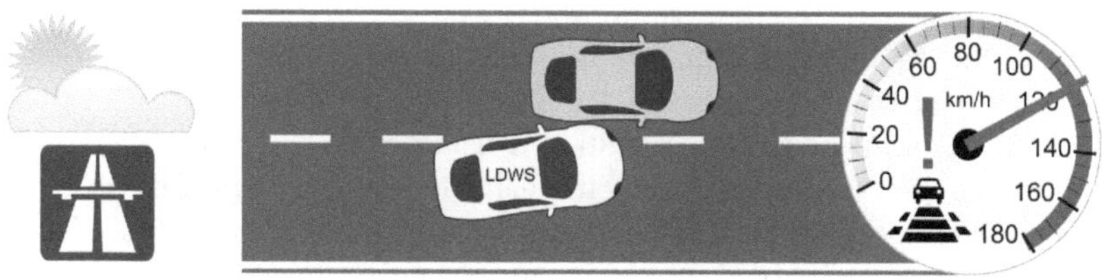

Fig. 2. Vehicle equipped with a lane departure warning system (LDWS) almost colliding with lateral traffic while driving at 120 km/h on a highway during daytime

1. *Requirements model:* The functional *requirements* of the system *component* (the LDWS) are formulated, e.g. the driver has to be alerted when the vehicle is leaving the traffic lane.
2. *Annotation:* The system component is annotated with domain attributes concerning the *operations* to be performed, the *environment* in which the system will be used, and the types of *accidents* which may occur due to malfunctions of the component. Regarding the LDWS, an example operation would be the continuous observation of the vehicle's lateral position within the traffic lane. The environmental conditions refer to driving situations on a normally frequented highway with a dedicated

road marking while driving with a velocity of 120 km/h during daytime in dry weather. They are annotated with the value for the probability of exposure (E) in this situation. Combining the component's operations with the environmental conditions automatically results in the respective *failure scenario*, e.g. a faulty detection of the driving lane on normally frequented highways while driving at 120 km/h during daytime in dry weather conditions. As system independent failure scenarios are referenced, the driver's controllability (C) in these scenarios can be specified independent of the system to be developed and prior to a risk analysis. The various types of possible *accidents* in the automotive domain are quantified via the potential severity (S) of the possible accident. As the LDWS observes the vehicle's position in the respective traffic lane, accidents resulting from unintentionally leaving the lane and colliding with surrounding traffic are possible in case of a potential component failure. Considering the failure scenarios and the possible accident types, the relevant *hazards* can be derived with the help of a generic hazard list that has been compiled in [1] and integrated into the DeSCAS ontology models. One of these hazards would be the possibility of undesired deviation from the traffic lane due to an error between set and actual value concerning the detection of the traffic lane.

3a. *ASIL classification:* Once all relevant system hazards have been identified, the risk class and thus the automotive safety integrity level (*ASIL*) of each hazard can be determined by means of the hazard analysis and risk assessment of the ISO 26262. For this purpose, the three parameters S (severity), C (controllability), and E (probability of exposure), which can be derived from the accident types, failure scenarios and environmental conditions linked to the respective hazards, are evaluated. The ASIL of all hazards is calculated using SWRL (Semantic Web Rule Language[4]). Overall, the safety integrity level of the component is determined within the DeSCAS ontologies by the highest ASIL of all identified hazards (e.g. *ASIL B*). This is accomplished by the OWL reasoner *Pellet*. There are four ASIL classes A, B, C, and D, where D represents the highest safety integrity level. A fifth class – QM (quality management) – does not impose any additional safety requirements on the system under development, but rather demands a regular quality management during the development process.

3b. *Safety requirements:* Applying the formalized ontology model of the ISO 26262 (see [7]), the calculated ASIL is further processed to infer traceability links to the ASIL-related *safety requirements* which themselves are associated with related *safety methods* and process phases and steps (i.e. *safety clauses*). In this case, a sample requirement of the system design phase would be the system design verification for compliance and completeness, which involves deductive analysis, highly recommended for ASIL B obligation.

[4] SWRL – http://www.w3.org/Submission/SWRL/

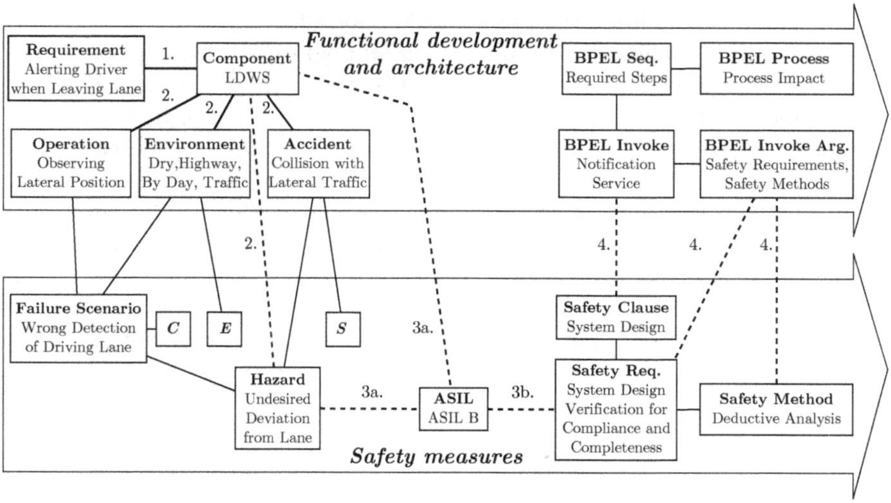

Fig. 3. A lane departure warning system is modeled with the design stream *functional development and architecture* (illustrated by the bold solid lines). Using inference techniques, product and process requirements can be reasoned automatically for the *safety measures* stream (see the bold dashed lines)

4. *Workflow model:* The inference on the formalized process model is used to derive a custom tailored workflow model for the individual developer in the development stream *functional development and architecture.* The tailored workflow model is further transformed into HTML documentation (more thoroughly described in [4]) and a BPEL (Business Process Execution Language[5]) workflow which can be instantiated. In the BPEL workflow, a *sequence* of notification services is *invoked* on the basis of derived ISO requirements and related methods. BPEL workflow monitoring can be used to track the progress of a workflow instance.

Regarding the implementation of these steps in a prototype toolchain, the transformation and reasoning steps have been implemented in Java, using the semantic reasoner *Pellet*[6] and the *OWL API*[7]. Toolchain integration is built on top of the *Apache Ant* build tool[8].

4 Lessons Learned from Applying Semantic Reasoning

On a conceptual level, OWL ontologies offer a convenient way for formally capturing and analyzing domain knowledge. The ability of defining descrip-

[5] BPEL – http://www.oasis-open.org/committees/wsbpel/
[6] Pellet – http://clarkparsia.com/pellet/
[7] OWL API – http://owlapi.sourceforge.net/
[8] Apache Ant – http://ant.apache.org/

tion logics axioms within OWL ontologies supports the automation and prototyping of analysis methods. Building upon an open-world assumption, OWL is very flexible and open so that one cannot rely on implicit assumptions. On a more technical level, formal reasoning on OWL ontologies may take a lot of computation time on a state-of-the-art personal computer due to the description logics axioms. To lower this time, the rule language *SWRL* and the query language *SPARQL* (SPARQL Protocol and RDF Query Language[9]) have been used in performance critical situations. Dividing major analysis steps into several individual steps, the computation time of reasoning can also be optimized. In terms of the DeSCAS ontologies, large ontologies with many axioms have been split up into smaller ontologies. However, this results in several consecutive reasoning steps.

5 Conclusion

To sum up, this paper has presented a methodology for interweaving differently geared development activities represented by the three development streams *functional development and architecture*, *safety measures*, and *human factors*, which are relevant within the automotive domain. The methodology forms a development proceeding for the design of safety critical automotive systems heavily relying on a formal base in contrast to most standard proceedings available. For this purpose, domain knowledge has been formalized using OWL ontologies illustrating how design decisions in one development stream may impact other domains and how this information can be used to reason about consequences of design decisions related to the current product development. However, formalization entails additional modeling effort when it comes to formalizing domain knowledge and standards, since a vast number of concepts have to be included in the OWL ontologies. On the other hand, once formalized domain knowledge and standards can be reused in other projects. The prototype toolchain of DeSCAS which has been used for the example lane departure warning system clarifies the advantages and disadvantages of applying OWL ontologies to logical reasoners for formal reasoning, and how to overcome problems arising from long computation times during reasoning.

Nevertheless, future research of DeSCAS will focus on extending and refining the proposed methodology by detailing the modeled domain knowledge (e.g. analyzing accident statistics to determine the severity of system hazards) and integrating more analysis methods (such as ASIL decomposition).

References

1. D. Beisel, C. Reuß, E. Schnieder, and U. Becker. Automotive Generic Hazard List. In *Automatisierungs-, Assistenzsysteme und eingebettete Systeme für Transportmittel (AAET)*, 2010.

[9] SPARQL – http://www.w3.org/TR/rdf-sparql-query/

2. J. Gačnik. Providing Guidance In An Interdisciplinary Model-Based Design Process. In *Proceedings of the 13th IEEE International Symposium on Object/component/service-oriented Real-time distributed Computing (ISORC 2010)*, Carmona, Spain, May 2010. IEEE Computer Society.
3. J. Gačnik, H. Jost, D. Beisel, J. Rataj, and F. Köster. DeSCAS Design Process Model for Automotive Systems – Development Streams and Ontologies. In *Safety-Critical Systems 2009*, number SP-2222 in Special Publications. SAE International, 2009.
4. J. Gačnik, H. Jost, F. Köster, J. Rataj, K. Lemmer, W. Damm, M. Fränzle, and E. Schnieder. DeSCAS – Formale Ontologien zur Verwebung von interdisziplinären Entwicklungsprozessen. In *AUTOMATION 2009*, number 2067 in VDI-Berichte. VDI Wissensforum GmbH, 2009.
5. ISO – International Organization for Standardization. ISO 17361: Intelligent transport systems – Lane departure warning systems – Performance requirements and test procedures, 2007.
6. ISO – International Organization for Standardization. ISO/DIS 26262: Road Vehicles – Functional Safety, December 2009. Draft International Standard.
7. H. Jost. Automating the Risk and Hazard Analysis via Generic Domain Concepts in Formal Ontologies. In *ESREL 2010, European Safety and Reliability Conference*, 2010.
8. J. Schwarz et al. RESPONSE 3 – Code of Practice for the Design and Evaluation of ADAS. In *PReVENT project deliverable D11.2*. Europe's Information Society, October 2006.

Calibration and Validation of Simulation Models for Investigation of Traffic Assistance Systems

Stefan Detering and Lars Schnieder

Institute for Traffic Safety and Automation Engineering
Technische Universität Braunschweig
38106 Braunschweig, Germany
{Detering, L.Schnieder}@iva.ing.tu-bs.de

Abstract. In recent years, initial proposals have been presented for advanced driver assistance systems (ADAS) which globally optimize traffic flow by means of the interaction of autonomous vehicles. This kind of ADAS will hereinafter be referred to as traffic assistance system (TAS). For the design, optimization and evaluation of these TAS, investigative simulations simultaneously considering both microscopic and macroscopic behavior are necessary. Therefore, in this paper a two-level approach for calibration and validation of traffic simulations is presented. This contribution presents a new measurement concept that is needed to gather the required data for the suggested two-level approach for calibration and validation. This concept advocates simultaneous data acquisition sourced from both a vehicle (microscopic) and an overall traffic (macroscopic) perspective which is furtheron compared to simulative data of both systemic levels. The paper describes the concept of calibration and validation of a car-following model with respect to intra- und inter-driver-variability, which makes it necessary to consider a distribution for each parameter in use. First empirical parameter distributions and their subsequent use in the Gipp's car following model are described in this paper. Results of a macroscopic validation with regard to headway distribution are presented. Compared to the current state of the art, the application of the two-level approach for calibration and validation with the gathered microscopic and macroscopic measurement data will enhance the possibilities to investigate the efficiency of TAS and yield results which are characterized by a higher degree of confidence.

Keywords: Advanced Driver Assistance System, Systems Theory, Traffic Modeling, Traffic Simulation, Calibration, Validation

1 Introduction

The current procedures of influencing traffic on highways as implemented at traffic control centers only affect limited sections of highways due to their intermittent mode of operation. Advanced driver assistance systems (ADAS),

E. Schnieder, G. Tarnai (eds.), *FORMS/FORMAT 2010*,
DOI 10.1007/978-3-642-14261-1_22, © Springer-Verlag Berlin Heidelberg 2011

which themselves use their measurements in order to continuously influence traffic flow by means of local interactions, could be used to remedy this deficiency. The primary objective for driver assistance systems so far is to increase either driving safety or driving comfort. As an enhancement, in recent years initial proposals have been presented for ADAS which are to optimize traffic flow globally by means of the interaction of autonomous vehicles without depending on a centralized traffic control center [7],[8]. These systems control the longitudinal vehicle behavior in order to optimize traffic flow and therefore can be considered as traffic assistance systems (TAS). For dimensioning and analysis of the effects of TAS on traffic flow simulative research is required.

In Section 2 the description of the system model is presented, which is applied to traffic modelling. Section 3 of this paper will show that investigations associated with TAS call for new requirements which have to be reflected. The review of the current state of the art in section 4 clearly shows that these requirements have not been considered in the last several decades as simulation investigations were designed with different objectives in mind. For the investigation of TAS the simulation model has to be valid on both the microscopic and macroscopic level. In order to reach this goal, it is necessary to perform a calibration and validation of the simulation model in use. Section 5 presents the new two-level approach for calibration and validation which aims to overcome the aforementioned deficiencies. Section 6 presents a data acquisition concept to obtain the necessary measurement data for fulfilling the determined requirements. Section 7 shows first results for calibration of the car-following behavior and the validation on macroscopic level for the headway distribution. The paper closes with a conclusion and an outlook for further research.

2 A system model to traffic modelling

A system can be described with its properties *state, function, structure* and *behavior* [10] which are interrelated specifically [11].

Furthermore systems can be characterized by an *abstraction hierarchy.* They are composed of a sum of parts which again can be decomposed into a sum of parts. Seen in detail those parts again show a certain complexity in terms of the system properties of *state, function, structure* and *behavior.* With reference to a specific level of abstraction a system has a superordinate and several subordinate systems. The particular system itself serves as a superordinate of subordinate system for other levels of abstraction. Following this principle of decomposition several layers of abstraction arise.

As soon as a system comes into existence by means of the combination of its parts, new properties emerge which have not been visible before and can not be explained by means of the properties of its isolated parts. This phenomenon is referred to as *emergence.* Emergence can be explained by means of the system structure. The elementary properties of systems previously discussed they can be further differentiated into *elementary properties* and *emergent properties*:

In the domain of traffic modelling different model structures developed, which may be classified with respect to different levels of emergence [3], [5], [6]. *Nanoscopic traffic models* simulate the cognitive processes within the driver or the mechanics of the vehicle. *Microscopic traffic models* simulate the behavior of single driver-vehicle units and describe their interactions by rule-bases that specify acceleration or velocity. Generally the dynamics are based on microscopic variables like distance or relative speed to the front or rear vehicle. *Macroscopic traffic models* neglect individual vehicles. They are devoted to aggregate state variables like traffic density and traffic flow for representing the collective behavior of vehicles. The equations are derived from the laws of nature and are structurally often similar to fluid dynamics.

3 Requirements for investigation of TAS

In recently developed TASs the local (individual) vehicle behavior is influenced in order to globally optimize traffic flow [7]. In order to obtain valid conclusion the following requirements for the investigation of TAS need to taken into consideration [2]:

For quantifying the benefit of these systems in real traffic a lot of test vehicles would be necessary. To equip so many vehicles is almost not possible. Therefore a *simulative approach* is necessary for the evaluation and optimization of these systems. Simulations have advantages relating to expenditure of time, costs and the possibiltiy to test different system alternatives.

The behavior of the driver-vehicle units is generally determined by the models in use, the interactions of the sub-models and especially by the parameters for the models. In order to obtain reliable simulation results, it is necessary to prove the validity of the selected parameters for each individual application [12]. Therefore, it is necessary to perform a *calibration*, i.e. adjusting the model parameters so that the simulation is sufficiently accurate when compared to actual behavior, as well as a *validation*, i.e. proving the model with simulation results obtained with the identified parameters which are compared to a second measurement data set. Quantitative statements can only be made from a model validated sufficiently.

Since the TAS changes the following behavior of a vehicle to its ahead-driving vehicle, it is absolutely necessary to use a traffic model for testing such a system which includes the headway behavior of the human driver as well as the TAS. Therefore in principle only a nanoscopic or microscopic simulation model can be used, which constitutes the *two-level approach* introduced in this paper. For simulating a lot of vehicles in a reasonable time span, nanoscopic simulation models are too complex. Therefore a simulation at microscopic level is necessary. A sole consideration of the microscopic level is not sufficient. Any adjustment of local microscopic traffic variables by the TAS emerges a certain global macroscopic behavior. In order to obtain reliable conclusions about the effect of TAS on macroscopic traffic behavior (in terms of traffic flow) the simulation needs to be calibrated and validated on both the microscopic and macroscopic level.

4 State of the art of calibration and validation

For the calibration and validation of microscopic simulation models, microscopic or macroscopic empirical data can be used. Therefore two different approaches can be identified:

- The first possibility is the calibration and validation of a microscopic simula-tion model with empirical microscopic data [1]. In this approach only isolated driver-vehicle units are considered. Therefore, the validation applies only to the microscopic level of the simulation. This approach does not explain the impact of the individual driver-vehicle unit behavior on the macroscopic va-riables (traffic flow, traffic density). That is the reason why this approach does not fulfill the requirements for the investigation of TAS.
- The second possibility is the calibration and validation of a microscopic simulation model with empirical macroscopic data [13]. Performing the validation only on the macroscopic level does not automatically result in a validation of the microscopic sub models. It may be possible that different combinations of microscopic behavior lead to the same macroscopic behavior. This approach does not guarantee the validity of the difference between human and TAS behavior. Therefore, this approach is not appropriate for the investigation of TAS either.

The review of currently available research shows that the microscopic and macroscopic levels of simulation currently co-exist and are not interwoven. In order to obtain meaningful simulation results for the optimization of TAS with a high degree of confidence, it is necessary to reflect both levels of simulation and to improve (calibration) and proove (validation) their faithfulness to reality. For this reason a two-level approach for calibration and validation is stipulated in the remainder of this paper.

5 Two-level approach for calibration and validation

The new proposed two-level approach is shown in Fig. 1. Calibration can be seen as a closed loop which involves the iterative execution of the following steps: First a microscopic simulation run is performed with default distributions of microscopic parameters given by the simulation tool. Subsequently, the microscopic empirical observed microscopic measurement variables like distance or relativ speed are compared against the results from the simulation and evaluated against a predefined threshold. In case the deviation between empirical results and simulation results is above the threshold in a third step, a thorough analysis of the underlying causes is needed. The fourth step is the model adjustment. This can either be an adjustment of the parameters (parameter variation), or a change in the underlying mathematical functions relating the parameters to each other (structural variation). After having adjusted the model we have to start with step one again and perform a new simulation run.

In the case the deviation during calibration is below the threshold, the validation of microscopic model parameters immediately follows the calibration. The calibrated model is transferred into a new, but comparable situation. For this new situtation a simulation run is performed. The microscopic simulation results are to be compared to an additional set of measured (microscopic) data. In case the previously defined threshold is exceeded a re-calibration is needed, thus the causal analysis is the subsequent step. The model gives valid results on the considered microscopic level when the previously determined error measurement bound is not exceeded.

As soon as microscopic parameters have been successfully validated a validation of macroscopic variables (e.g. traffic flow, traffic density, mean speed) is possible. First a microscopic simulation run is performed. In a second step the macroscopic simulation results are compared against empirical data. In case the results stay within the permissible range the calibration and validation on microscopic and macroscopic level has been successful. Otherwise a re-calibration on microscopic level becomes necessary. Now it becomes obvious that calibration is only possible on the microscopic level (as microscopic variables are independent variables whereas macroscopic variables are dependent variables).

6 Data acquisition concept

To carry out the two-level calibration and validation both macroscopic and microscopic real-traffic variables need to be gathered in such a manner that they share the same time and geographical reference. Additionally, the equivalent microscopic and macroscopic simulation results are required.

As input for simulation and for validation on macroscopic level an *empirical data acquisition on macroscopic level* is necessary. The knowledge about the traffic flow within the considered section of the highway is essential. Along the track this data is available at traffic control centers. However, the available data is insufficient and needs to be amended with empirical data from on- and off-ramp traffic flow. For this reason a use of additional (temporary) sensors is proposed. For an installation nearby the track the radar sensor as well as the passive infrared sensor is suitable [9].

To describe the individual driver behavior, *empirical data acquisition on the micro-scopic level* is necessary. The current focus of investigation of individual driver behavior is the car-following behavior. In the selected approach an experimental vehicle is equipped with sensor systems based on radar and lidar technology, which measure distance and relative speed to the preced-

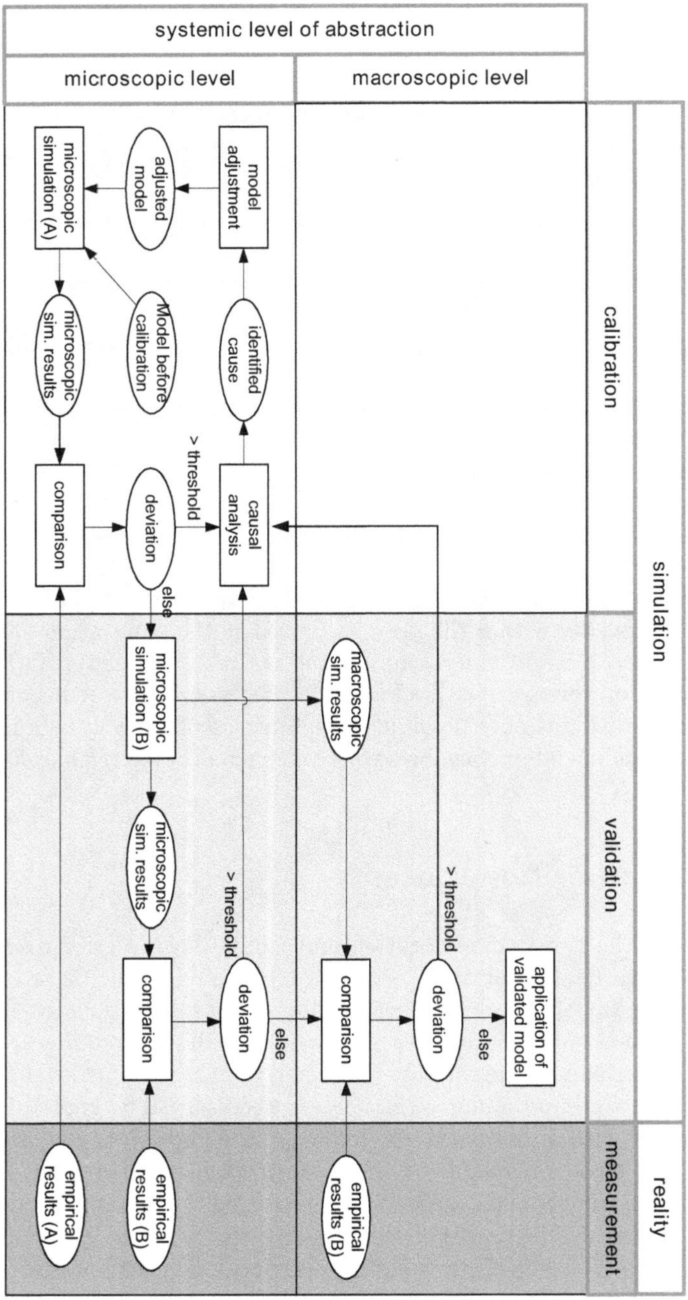

Fig. 1. Calibration of a microscopic simulation model on microscopic level and validation on two systemic levels of abstraction

ing (used for the investigation of *intra-driver-variability*) and the following (used for the investigation of *inter-driver-variability*) vehicle. These data are recorded together with the speed of the experimental vehicle. So far only one vehicle has been equipped. This vehicle can be used to prove the possibility of measuring the necessary required variables. The goal is to equip several vehicles with this sensor system in order to perform measurements with all these vehicles on a predetermined section at the same time. In a first step, with the one equipped vehicle 45 test runs with altogether more than 3000 km on the Autobahn A2 between Braunschweig and Hanover have been carried out until now.

To compare the microscopic and macroscopic empirical data sets with the simulation results, the simulation tool must offer the possibility to obtain *simulation data on microscopic and macroscopic level.* At the moment, the comparison is implemented by the simulation tool AIMSUN. This simulation offers an established microscopic car-following and lane-changing model as well as a detailed description of these models. The measurement of macroscopic data is by default possible with the simulation tool. The measurement of microscopic data cannot be accomplished by the standard simulation tool. Hence, the application interface of AIMSUN is used to extend the functionality and to obtain microscopic measurement data from individual vehicles which can be compared to the empirically observed behavior.

7 Empirical Results

The implementation of the entire data acquisition concept is currently under development. This section shows first results for the data acquisition and calibration of the car-following behavior and its validation on macroscopic level for the headway distribution.

For calibration the Gipps car-following model was applied [4]. A genetic algorithm and a goodness-of-fit measure based on Theil's U considering speed and distance was used in order find a proper approximation. In this paper the results for the parameter b, which describes the most severe braking that the driver wishes to undertake are presented.

In a first step for the analysis of *intra-driver-variability* of car-following-behavior a total of 79 car-following trajectories from the equipped vehicle as a following vehicle were used for calibration. The test driver shows different values for parameter b for different car-following trajectories, which can be interpretated as an *intra-driver-variability.* Additionally the parameter values for speeds of the equipped vehicle lower than 20 m/s are almost all greater than $2m/sec^2$, for speeds greater than 20 m/s the parameter values are all less than $2m/sec^2$. These two results show the *intra-driver-variability* and a speed-dependency of car-following behavior. Amazingly, many previous investigations regarding car-following behavior have only considered speeds up to 20 m/s and therefore could not investigate this speed-dependency of driver behavior [1], [7].

In order to analyze *inter-driver-variability* of parameters used to describe car-following behavior in traffic simulations a total of 85 car-following trajectories from the vehicles, following the equipped vehicle were used for calibration. In this case the identified values for parameter b indicate that those drivers are inclined to exert a more severe braking (apply higher rates of deceleration) compared to the test driver of the equipped vehicle. Smaller values for parameter b are due to the smaller distance the following vehicles kept to the equipped vehicle. Because the driver of the equipped vehicle knows that he takes part in an investigation, it can be assumed that the test driver was driving more carefully than the following vehicles. This reactivity is also known as the so called "Hawthorne effect". The identified parameter values are distributed from $2m/sec^2$ to $8m/sec^2$ indicating the *inter-driver-variability.*

In a second step a comparison of aggregated empirical distance behavior (expressed in terms of the time headway in seconds) of the vehicles following the equipped vehicle, the simulation results using the standard parameter set of AIMSUN and the simulation results using the identified parameter set for car-following behavior identified for *inter-driver-variability* has been performed. An adjustment of the identified parameters was necessary as AIMSUN assumes that parameters are distributed normally, but some identified parameter distributions were skewed to the right. For all other parameters (not considering car-following) the default parameters of AIMSUN were used.

In the case using the standard parameter set of AIMSUN it becomes obvious that the mean value differs from empirical data and also the standard deviation is lower than the one derived from empirical data. This comparison demonstrates the need to calibrate and validate the simulation model. In the case using the identified parameters for the car-following model the time gaps still differ from the empirical data. In reality, shorter headways could be observed. However, in comparison to the standard parameters a closer approximation of simulative data with empirical results could be achieved. Particularly the *inter-driver-variability* expressed by the larger standard deviation is represented much more precisely. The systematic weakness of the simulation tool seems to be the main reason for the remaining deviation. Thus, the adaption of the identified parameters for the car-following model in AIMSUN mentioned above was necessary.

8 Conclusion

It has become obvious that previous approaches towards a calibration and validation of traffic simulation models do not serve the needs for investigation and optimization of TAS. Therefore a new two-level approach was presented, which combines a calibration on the microscopic level with a validation on the microscopic and macroscopic level. The paper presented a data acquisition method, which allows the consideration of measurement data of the microscopic and macroscopic level originating from the same time span and the same road network. To obtain the empirical microscopic data a vehicle was equipped with the necessary measurement instruments. First results from the calibration of the car-following behavior have been presented showing both *intra- and inter-driver variability* as well as the speed dependency of parameter values. The next steps are the investigation of lane-changing behavior and the realization of the entire data acquisition concept using several individual vehicles and additional local sensor systems. In the future the approach set out in this paper allows for a more accurate and precise design and comparison of TAS.

References

1. Elmar Brockfeld, Rene Kelpin, and Peter Wagner. Performance of car following behaviour in microscopic traffic flow models. In W. Möhlenbrink, editor, *2nd International Symposium "Networks for Mobility"*, pages 43 – 43. Universität Stuttgart, 2004.
2. S. Detering and E. Schnieder. Requirements for precise simulation models for traffic flow optimizing adas. In *12th IFAC Symposium on Control in Transportation Systems (CTS'09)*, Redondo Beach, USA, 2009.
3. Stefan Detering. *Verkehrsleittechnik - Automatisierung des Straßen- und Schienenverkehrs*, chapter Flusssteuerung im Straßenverkehr, pages 155 – 194. Springer, 2007.
4. Peter G. Gipps. A behavioural car-following model for computer simulation. *Transportation Research Part B: Methodological*, 15(2):105–111, April 1981.
5. Dirk Helbing. *Verkehrsdynamik - Neue pyhsikalische Modellierungskonzepte.* Springer, 1997.
6. B.S. Kerner. *Physics of Traffic: Empirical Freeway Pattern Features, Engineering Applications, and Theory.* Springer Verlag, 2004.
7. Arne Kesting. *Microscopic Modeling of Human and Automated Driving: Towards Traffic-Adaptive Cruise Control.* PhD thesis, Technische Universität Dresden, Fakultät für Verkehrswissenschaften "Friedrich List", 2008.
8. Florian Kranke, Holger Poppe, Arne Kesting, and Martin Treiber. Der Baustellenlotse - Ein stauvermeidenes Fahrerassistenzsystem. *Straßenverkehrstechnik*, 1:12–19, 2010.
9. Timo Schneider. Sensorsysteme zur Verkehrslageerfassung auf Bundesautobahnen. Bachelorarbeit, Institut für Verkehrssicherheit und Automatisierungstechnik, Technische Universität Braunschweig, 2010.
10. Eckehard Schnieder. *Methoden der Automatisierungstechnik - Beschreibungsmittel, Modellkonzepte und Werkzeuge für Automatisierungssysteme.* Vieweg, 1999.
11. L. Schnieder. Towards terminological rigour in the specification of complex automation systems. In G. Schnieder, E.; Tarnai, editor, *Proceedings of Symposium FORMS/FORMAT*, pages 141–148, Budapest/Hungary, 2008.
12. Roland Trapp. *Hinweise zur mikroskopischen Verkehrsflusssimulation.* FGSV Verlag, 2006.
13. Michael Zhang, Ma Jingtao, and Hu Dong. Developing calibration tools for microscopic traffic simulation final report part 2: Calibration framework and calibration of local/global driving behavior and departure/route choice model parameters. Technical report, California Path Program, Institute of Transportation Studies, University of California, Berkely, 2008.

Model-based Integration Framework for Development and Testing Tool-chains *

B. Polgár, I. Ráth, and I. Majzik

Department of Measurement and Information Systems,
Budapest University of Technology and Economics
{polgar,rath,majzik}@mit.bme.hu

Abstract. System development processes are typically supported by dozens of different tools that assist the designer in various phases of development like modeling, verification, source code generation, testing. Tool-chains can be formed by the integration of tools that are related to the subsequent steps of the process. In this paper, we present a service-oriented, metamodel-driven, process-centric approach for the definition and execution of these tool-chains. Related data are handled as an important part of the process as the traceability of these is needed for the certification of the systems. The implementation is provided as an open, extensible framework. The approach is demonstrated using a model based test case generation process applied for automotive and railway systems.

Keywords: Tool Integration, Execution Support for Tool-chains, Process-centric, Service-oriented, Model-driven Support of Development and Test Processes

1 Introduction

Motivation. During system development — especially in the development of safety-critical systems, e.g., in the field of automotive, avionics or railway — several tools are used for the different aspects, i.e., for modeling, transformation, verification, testing, analysis and code generation. We faced the challenge in several of our projects to integrate these tools. Our efforts are concentrated on designing and implementing the integrated end-to-end design tools with transparent transformations which are used to automatically map the design and specification models to analysis and validation domains for a thorough verification and validation process.

The goal of the MOGENTES project [1] is to significantly enhance testing and verification of dependable embedded systems by means of automated generation of efficient test cases from engineering models. In the project the

* This work was partially supported by the EC (FP7-STREP MOGENTES) and by the National Office for Research and Technology (OMFB 1316/2009). Special thanks to I. Ágoston and G. Juhász for implementing parts of the framework.

E. Schnieder, G. Tarnai (eds.), *FORMS/FORMAT 2010*,
DOI 10.1007/978-3-642-14261-1_23, © Springer-Verlag Berlin Heidelberg 2011

newly developed test generation technologies should be integrated with existing modeling tools and with the test environment of the industrial partners from the automotive and railway domain. The integration should be realized in a seamless way, i.e., domain experts with limited knowledge and experience in usage of formal methods (that will be used for test generation) should also be able to use them with minimal learning effort.

Related Work. Tool integration has been a hot research topic over the past years. There have been a number of early attempts (such as [2]) at integrating development tools within the context of a well-defined process. They were rather suited for a particular type of application rather than being generally usable.

The experience from these approaches has been collected and synthesized into design patterns [3] which are common to most tool integration approaches of today. An important class is *metamodel-driven tool integration* [4], which is based on the idea of a model bus, a data repository which captures semantic information on the data that is exchanged between the tools and provides uniform persistence support. Recent initiatives ([5]) target advanced features such as model difference computation and model merging, but their scalability to industrial model sizes is yet to be evaluated. The workflow-based approach [6] has been (partially) implemented in a number of tools. The SENSORIA Development Environment [7] offers Eclipse-based integration interfaces and a simple orchestration language in which small tool integration processes can be described. jETI [8], a similar tool integration framework, is targeted at remote invocations for Eclipse-based tools using Web Services technology. With the increasing emphasis on organized collaborative work in software development, high-level team management tools such as Rational Jazz [9] are emerging, driven by precise process models exported from modeling environments like the Rational Method Composer or the Eclipse Process Framework Composer [10].

Goals. In our solution our goal was to reuse and combine existing technologies and extending them only if necessary. Thus, our approach is targeted as a complementary contribution to high-level collaboration integration environments. We apply both the metamodel-driven and the process-driven tool integration patterns and with our solution we address the following objectives:

- **Model-based construction of tool-chains.** A process metamodel is created for the definition of those pieces of the development processes that are related to the execution of tools. The integration of tools required for a given activity is specified using a *process model* that is an instance of this metamodel. It includes the tasks, the supporting tools, and also the input/output artefacts. Available tools can be organized into libraries.
- **Automatic derivation of execution configuration.** The process model is mapped automatically to the input language of a *process execution*

engine that invokes the tools specified in the process model and ensures the proper handling of the related artefacts. This way the process developer is able to focus on the semantics of the tool-chain without dealing with specific lower level notations and configuration options of the execution engine.
- **Model-based execution and supervision of tool-chains.** A management interface is provided that uses the platform specific model for the execution of the tool-chains, and the platform independent model as the interface to the system developer who executes the process.
- **Data integration.** The artefacts processed and generated by a tool-chain are handled as an integral part of the process: these are represented explicitly in the models, are stored in persistent data repositories, and the download and upload are managed automatically.
- **Definition of a flexible framework.** In the architecture key components are identified independently of the underlying technology. For the implementation of the components high level, modern technologies having strong industrial support are used if possible. Between the components interfaces are defined in order to be able to change the implementation technology.
- **Traceability of processes.** Certification of the developed system is a need in case of safety-critical systems, thus all steps executed during the development process with all the data handled shall be documented and be able to traced back upon request.

The paper is organized as follows: the architecture of the framework with key components is described in Section 2. In the next section details about the implementation are provided, while Sec. 4 demonstrates the concepts in a model based test case generation case study. Finally we conclude our work.

2 Architecture of the Tool Integration Framework

Based on our research and development experience and on the related works in the field we have identified the key components that are needed to fulfil the goals described in the introduction. The architecture is depicted in Fig. 1. The components can be grouped into four categories: process modeling, tool management, data management, and process execution.

Tool Management. A *Tool* represents an executable program that performs one or more tasks during the development. Tools are independent of the framework and can be implemented in different languages and according to different technologies (e.g., also as web service). The only requirement is to have the functionality exposed as a well-defined interface which is externally accessible.

For each tool a *Connector* shall be implemented which exposes the functionality of the tools as services. Connectors are usually simple wrapper classes that provide homogeneous programming interface and hide the heterogeneity of the tools. Connectors shall be registered to the *Tool Manager* which composes a runtime *Repository* and makes accessible the tools and their services.

Data Management. During development another important part is the data on which the different tools are operating. These can come from multiple sources: e.g., from local or remote file systems, from databases, or from version handling repositories. The *Artefact Manager* provides data management related services to other components of the framework. It is composed of a general component and multiple repository specific data connectors. A uniform address structure is used to address the data in different repositories.

Process Modeling. Process Models are composed of services and related data: it contains the information about what is done on which artefact and with which tool. Both control flow and data flow appears in the model. The control flow determines the sequence of tasks, while the data flow determines the relation between tools and data.

First, a platform independent process model is created which is independent of the execution environment. This contains the information *"what to do"*. If an execution engine is selected it shall be transformed to the format needed by the engine. This platform specific model also includes the information *"how to do"*.

Process Execution. The *Process Execution User Interface* can be used to deploy the platform specific process models to the *Process Execution Engine*, and to start and trace the execution of these. When execution is initiated (i)

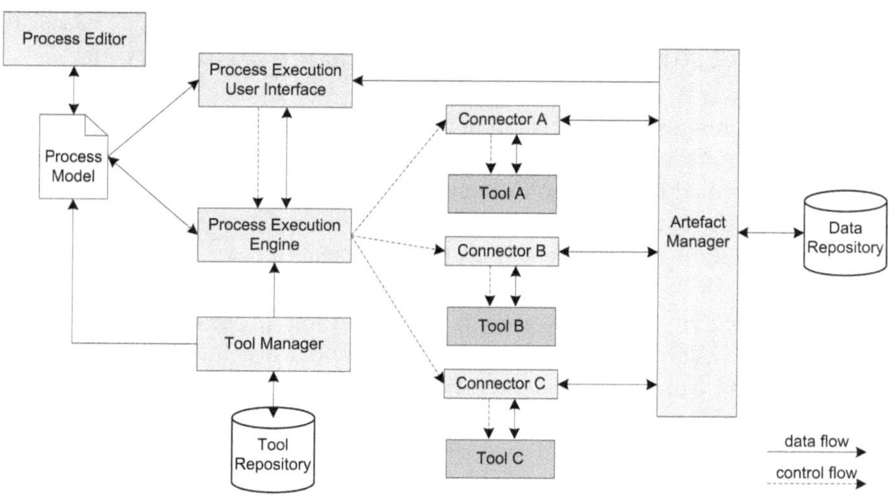

Fig. 1. Framework architecture

tools referenced in the model are retrieved from the Tool Manager and the needed service is invoked through the Connector, (ii) input data referenced in the model is downloaded automatically from the repository and provided to the service, and (iii) output data is uploaded to the repository. The User Interface contains the image of the platform independent version of the process model, where the state of the execution and the related data can be traced.

3 Implementation of the Tool Integration Framework

The tool integration framework is implemented as an Eclipse based tool and the standard Eclipse extension mechanism is used for the extendability of the framework.

Tool Management. The Tool Manager component was developed in the Sensoria project and it is called SDE [7]. This provides a *tool extension point*, to where the tool Connectors shall register. A local tool repository is composed of this registered tools. If the Tool Manager is deployed on multiple hosts, it is possible to add the remote tool repositories to the local one and access the tools registered there through r-OSGi communication. This way the framework can be used in a distributed environment.

Data Management. The general component of the Artefact Manager provides a *repository extension point*, to where the repository connectors can register. Currently connectors are implemented for the file system (as this is handled also as a data repository), for the Java Content Repositories (the Apache Jackrabbit reference implementation is used [11]), and for the Subversion repository [12].

Process Modeling. For the creation of process models an Eclipse GMF [13] based graphical editor is developed. Here the process can be composed of the services deployed on the local or on a remote tool repository, and artefacts contained by the available data repositories can be referenced.

Process Execution. The JBoss jBPM workflow engine [14] is integrated as the process execution component of the framework. It provides a persistent store for the processes, where information about all executions are stored. The process execution user interface part of the framework provides a catalogue of the deployed and instantiated processes, makes possible to deploy and instantiate new processes and to trace the state of the selected process.

jPDL (jBPM Process Definition Language, [15]) is the native process format of jBPM, this platform specific process model is transformed automatically from the "general" process model created in the process editor. We have extended the jPDL language with the notion of data nodes, and provided action handlers for the tool and data nodes. The former invokes the referenced service on the local or remote host, while the later transparently downloads input data of the tools from repositories or uploads output data to the repositories as determined by the process model.

4 Case Study: Model-based Test Case Generation

Description of the example. The example is taken from the MOGENTES project [1], where the goal is the model-based generation of efficient test cases. Demonstrators are from the field of automotive and railway, thus their systems are real-time and safety-critical, and needs intensive testing. For the formal specification of the model UML models are used with a precise semantics. The structural parts are described by class diagrams, and the behavior of the objects by state machines. During testing, the test goal is to verify the fulfillment of the requirements related to the behavior, which are associated to elements of the UML state machines. As real-time models are considered, the state machines contain also time related transitions, and their semantics is provided through a mapping to timed automata [16].

The test case generation process. It is a common practice to use model checkers for generating test cases for state based systems. There is a COTS tool environment UPPAAL [17] for timed automata, thus we have transformed UML models to timed automata, generated traces with UPPAAL that can be used as test cases, and converted the traces to abstract test cases (ATC) which have a pre-defined format and in which the concepts of the original UML models are referenced. The transformation from UML to timed automata is performed in two steps: first the hierarchical state machines are flattened to Kripke-structures (so called SMTE models), and those are mapped to the input language of UPPAAL. During the transformation a mapping file between the concepts of UML and UPPAAL is also generated, which is used during the conversion of UPPAAL traces to ATCs. The process is depicted in Fig. 2.

The deployment of the framework and the related components. During the implementation three tools are developed that provides four services: a transformation tool with two transformations (UML-to-SMTE, SMTE-to-UPPAAL), a UPPAAL based tool generating a trace from a model if a test goal is pro-

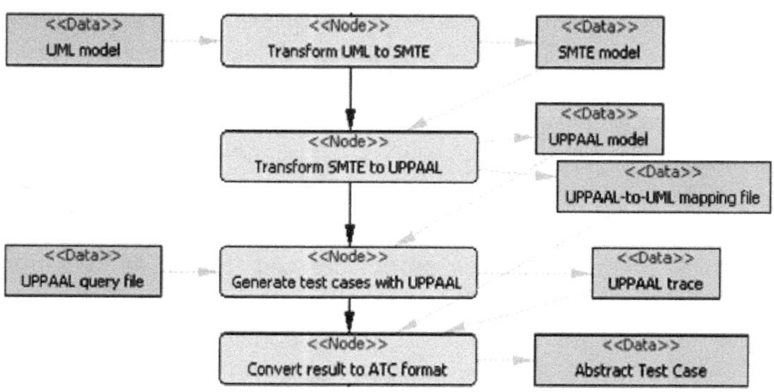

Fig. 2. The process of timed automata based test case generation

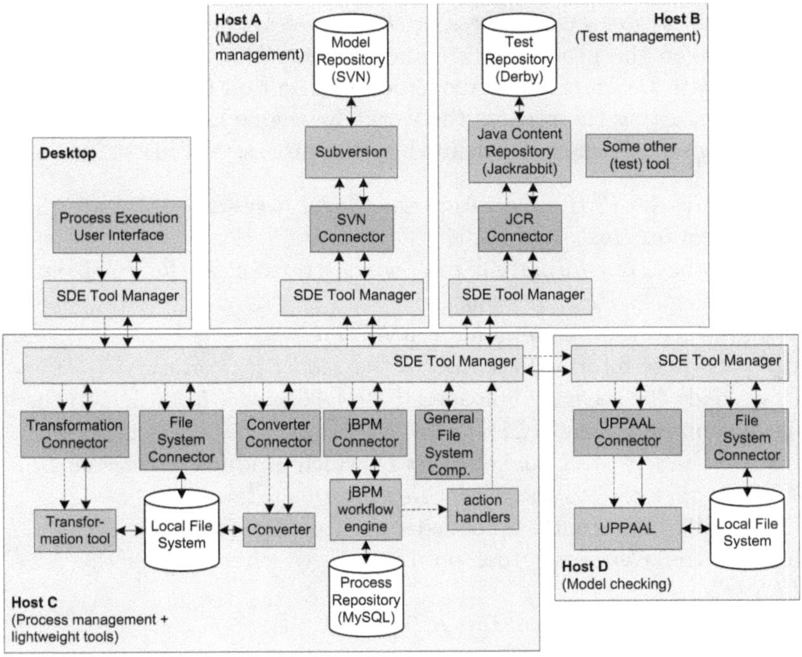

Fig. 3. The deployed components of the example system

vided, and a converter tool producing an ATC from a trace using the UML-UPPAAL mapping information. For all the three tools framework connectors are provided.

Let's assume that the railway company –who wants to generate test cases– stores their models in an SVN repository deployed on *Host A* (where the UML model is located and to where the intermediate models and results shall be uploaded), and has a test server (*Host B*) with a Jackrabbit JCR based test repository on it (to where the ATC shall be uploaded). To use our framework, we need a process server (the jBPM workflow engine), which is worth to be deployed in a separate machine (*Host C*). In our example, we chose to deploy the transformation and the converter tools also to *Host C* as these are quite lightweight tools, but the UPPAAL model checker is deployed to a separate, strong machine (*Host D*) as in case of large models it needs more resources (Fig. 3).

To allow the process server to access the tools and data repositories, the *SDE Tool Manager* component shall be deployed to all hosts with the appropriate tool or repository connectors. In addition, the *General Data Repository Component* shall be deployed to the host where the process execution engine is located (*Host C*), and the *File System Connector* to the hosts where tools operating on local files are deployed (*Host C, D*).

The execution of the TCG process in the framework The deployment of the TCG process to the process server and the execution of it can be initiated from a separate *Desktop*, where the Process Execution User Interface is available. When starting the process the workflow engine executes automatically the following steps (which are defined by the process model):

- Downloads the *UML model* from the *Model Repository* (*MR*) to the local file system on *Host C* (*LFS_ C*).
- Invokes the *Transformation tool*, which transforms the *UML model* to *SMTE* then to *UPPAAL model*,
  and produces the *mapping file* (on *LFS_ C*).
- Uploads these intermediate files to the *Model Repository*.
- Downloads the recently uploaded *UPPAAL model* from *MR* to the local file system on *Host D* (*LFS_ D*).
- Invokes the *UPPAAL tool* on *Host D*, which produces a *trace* on *LFS_ D*.
- Uploads the *trace* to the *Model Repository*.
- Downloads the recently uploaded *trace* from *MR* to *LFS_ C*.
- Invokes the *Converter* tool on *Host C*, which produces the *ATC* on *LFS_ C*.
- Uploads the *ATC* to the *Test Repository*.

During the execution the status is updated on the user interface on the Desktop. At the end all related information is stored in one of the repositories in a traceable manner, which also supports the certification of the development process and the developed application.

5 Conclusions

We have presented a general purpose tool integration framework composed of widely supported and used technologies and tools, which enables the automated execution of tool-chains used during the development of safety-critical systems, where dozens of dependent tools are used. The unnecessary details of the execution of e.g. verification, test case generation or test execution processes are hidden from the engineer, who can concentrate on the development of the application and consider only the result of the verification or the testing as part of the application refinement.

The framework can be operated in a distributed environment as illustrated in the test case generation example. It also supports the traceability of the artefacts handled during the process in a seamless way enabling the certification of the development process and the developed application.

References

1. The MOGENTES Project: Model-based Generation of Tests for Dependable Embedded Systems, EU FP7 Research Project http://mogentes.eu.

2. Einar W. Karlsen: The UniForM WorkBench: A Higher Order Tool Integration Framework. Lecture Notes In Computer Science **1641/1999** (1999) 266–280.
3. Gábor Karsai, András Lang, Sandeep Neema: Design patterns for open tool integration. Software and Systems Modeling **4**(2) (2004) 157–170.
4. C. Amelunxen, F. Klar, et al: Metamodel-based Tool Integration with MOFLON. In: Int. Conf. on Software Engineering, ACM (2008) 807–810.
5. R. Salay, et al: An Eclipse-based Tool Framework for Software Model Management. In: OOPSLA Workshop on Eclipse Technology eXchange, ACM (2007) 55–59.
6. F. Corradini, L. Mariani, E. Merelli: An agent-based approach for tool integration. Int. Journal on Software Tools for Technology Transfer **6**(3) (2004) 231–244.
7. The SENSORIA Project: The SENSORIA Development Environment Homepage (2009) `http://svn.pst.ifi.lmu.de/trac/sct`.
8. Tiziana Margaria, Ralf Nagel, Bernhard Steffen: jETI: A Tool for Remote Tool Integration. Lecture Notes in Computer Science **2440/2005** (2005) 557–562.
9. IBM Rational: Jazz Community Site `http://jazz.net/`.
10. P. Haumer: Increasing Development Knowledge with EPF Composer. Eclipse Review (2006) `http://haumer.net/rational/publications.html`.
11. Apache Jackrabbit: The reference JSR-170 implementation `http://jackrabbit.apache.org/`.
12. Apache Subversion: A version control system `http://subversion.apache.org/`.
13. GMF: Eclipse graphical modeling framework `http://www.eclipse.org/modeling/gmp/?project=gmf`.
14. John Koenig: JBoss jBPM White Paper. Technical report, The JBoss Group / Riseforth.com (2004) `http://jbossgroup.com/pdf/jbpm\_whitepaper.pdf`.
15. JBoss.org: jPDL User Guide. (2006) `http://www.jboss.org/jbossjbpm/jbpm\_documentation`.
16. G. Pintér, I. Majzik: Test Case Generation Base on an Integrated UML Data and Behavior Semantics. In: submitted to FORMS/FORMAT 2010. (2010)
17. Larsen, K.G., Pettersson, P., Yi, W.: UPPAAL in a Nutshell. International Journal on Software Tools for Technology Transfer **1**(1–2) (October 1997) 134–152

Automatically Deriving Symbolic Invariants for PLC Programs Written in IL*

Sebastian Biallas[1], Jörg Brauer[1], Stefan Kowalewski[1], and Bastian Schlich[2]

[1] Embedded Software Laboratory, RWTH Aachen University
{biallas, brauer, kowalewski}@embedded.rwth-aachen.de
[2] Industrial Software Systems, ABB Corporate Research
bastian.schlich@de.abb.com

Abstract. In this paper, we propose a new approach to automatically derive invariants from Programmable Logic Controller programs by symbolically rewriting Instruction List code. These invariants describe the relations between all variables and capture the behavior of the program. Usually, invariants are created by users and verified using formal verification techniques such as model checking or static analysis. The process of manually deriving invariants, however, is error-prone and lengthy. Our approach generates these invariants automatically and removes the need to use formal verification techniques to verify them. Users only need to inspect the generated invariants and compare them to the expected program behavior. Using three example programs of different sizes, we show that the generated invariants are easy to understand and that the approach indeed scales for larger programs.

Keywords: Program Verification, Invariants, PLC, Instruction List

1 Introduction

Programmable Logic Controllers (PLCs) are frequently used in safety-critical systems, where the application of formal verification methods is recommended [10]. In the past, formal methods such as model checking [17,15] or static analysis [8] have been applied to this task. Model checking, for instance, is used to verify whether the model of a systems satisfies the system's requirements. The generation of the model can in many cases be done automatically. However, expressing requirements, which are often given in natural language, in terms of temporal logic, is a very time-consuming and error-prone process.

Despite the advances in this regard [12], this drawback limits the applicability of formal verification methods to industrial applications. To alleviate

* The work of Sebastian Biallas was supported by the DFG. The work of Jörg Brauer and Stefan Kowalewski was, in part, supported by the DFG Cluster of Excellence on Ultra-high Speed Information and Communication (UMIC), German Research Foundation grant DFG EXC 89.

E. Schnieder, G. Tarnai (eds.), *FORMS/FORMAT 2010*,
DOI 10.1007/978-3-642-14261-1_24, © Springer-Verlag Berlin Heidelberg 2011

this problem, we propose to use a different approach: Instead of verifying a previously stated specification, our method derives symbolic invariants.

Invariants as such have long served for reasoning about correctness of programs. They can describe relations between the valuations of variables that hold after the execution of a program fragment regardless of the input values [7]. Examples of invariants are, for instance, $x = -y$ or $x \leq 0$.

To derive invariants, our approach considers an input program written in Instruction List (IL) and iteratively rewrites this program into a symbolic representation over quantifier-free linear arithmetic with Boolean connectives, following the semantics of the involved operations.

Our method is different from existing work in that it directly targets PLCs running in the *cyclic execution mode*, which consists of three steps, each of which is executed atomically: reading inputs, processing data, writing outputs. Our approach translates the semantics of the instruction set into linear constraints and then uses a set of rewriting rules to derive invariants. Since PLCs are typically running in safety-critical systems where timeliness is a strong concern, most of these programs do not consist of long running loops (not to say, infinite loops). This property ensures that the derived arithmetic expressions do not grow without bounds.

In previous work, Pavlovic et al. [15] have translated programs written in *Statement List* into the input language of the model checker NuSMV [4]. Their most important contribution was the formal verification of the program depicted in the left hand side of Fig. 1. As input for NuSMV, they used the following specification given in the temporal logic LTL [6]:

$$G(\texttt{PC} = 2 \Rightarrow \texttt{Byte} = (\texttt{Bit0} + 2 * \texttt{Bit1} + 4 * \texttt{Bit2} + 8 * \texttt{Bit3} \\ +16 * \texttt{Bit4} + 32 * \texttt{Bit5} + 64 * \texttt{Bit6} + 128 * \texttt{Bit7})) \quad (1)$$

In summary, this specification states that the program converts a bit-vector of length 8 into an unsigned byte. The approach described by Pavlovic et al., however, has two drawbacks. On the one hand, the runtime requirements for their approach is significant: Model checking took approximately 8h (even though this could be reduced to 113s with manual intervention). On the other hand, the specification needs to be formulated manually.

Our method can be seen as a response to these problems: By rewriting the instructions in the program, it derives the stated invariant automatically. Further, the runtime is essentially non-measurable, requiring less than 0.1s overall. In summary, we make the following contributions:

- We detail an automatic approach for deriving invariants from PLC programs written in IL.
- We present the effectiveness of this approach on three examples, where precise and expressive invariants were derived.

Our approach is as follows. The program is first rewritten into a so-called static single assignment (SSA) form, that explicitly shows all calculation steps in a symbolic representation (cp. Sect. 2). Then, this explicit form is used to derive symbolic invariants by analyzing the SSA expressions stored in output variables at the end of the PLC cycle (cp. Sect. 3). We additionally present a second example, where two invariants are derived depending on the actual input values and a third example analyzing an implementation of a PLCopen safety function block. The paper is concluded by presenting related work (cp. Sect. 4) and discussing results and future work (cp. Sect. 5).

2 Rewriting of IL Programs Into SSA Form

For the application of our method, we are analyzing IL programs, which is is one of the standardized languages for programming PLCs [9]. It is accumulator-based and similar to many machine languages. With its simple semantics it is ideal for deriving symbolic information.

We will motivate our approach with the `FromByte` program according to Pavlovic et al. [15], which was translated to IL [17]. An excerpt of this program is shown in Fig. 1 on the left side. It has eight Boolean inputs named `in0` to `in7` and converts these to the byte represented by them. This is accomplished by converting the inputs to the corresponding integer (*false* to 0 and *true* to 1), multiplying them by their significance and adding up the results in a temporary variable called `temp`.

For deriving symbolic invariants, we begin by rewriting IL programs into an SSA form [5]. In this form, each IL instruction is written as an assignment. Each of these assignments creates a new instance of the accumulator or a variable, indicated by a superscript number. If, e. g., the current accumulator $\mathsf{acc}^{(i)}$ is incremented by 1, the SSA expression

$$\mathsf{acc}^{(i+1)} := \mathsf{acc}^{(i)} + 1$$

would be generated. The superscript number is called *instance number.* On the left hand side (LHS) of such expressions, there is either a new instance of the accumulator or a new instance of some program variable. The right hand is side (RHS) is either

- a constant,
- a program variable that was not yet used on a LHS,
- an arithmetic, logic or relational operation of existing LHSs, or
- a data type cast. For example, a cast of the accumulator to the `BYTE` type is written u8(acc), indicating the 8 bit unsigned type.

The transformation of the IL program into SSA form is performed automatically. In order to achieve this, different execution paths are separated by partitioning the possible ranges of input variables as presented in [17]. Additionally, all loops are unrolled, removing conditional execution all together. Formally, the translation is shown in Fig. 2 for the instructions `LD`, `ST`, `ADD`, `SUB` and `AND`. We assume that there are already i instances of the accumulator, so a write to the accumulator creates instance $i + 1$. For the store to

1	`LD`	`in0`	$\mathsf{acc}^{(0)} := \mathit{in_0}^{(0)}$
2	`BOOL_TO_BYTE`		$\mathsf{acc}^{(1)} := \mathrm{u8}(\mathsf{acc}^{(0)})$
3	`ST`	`temp`	$\mathit{temp}^{(0)} := \mathrm{u8}(\mathsf{acc}^{(1)})$
4	`LD`	`in1`	$\mathsf{acc}^{(2)} := \mathit{in_1}^{(0)}$
5	`BOOL_TO_BYTE`		$\mathsf{acc}^{(3)} := \mathrm{u8}(\mathsf{acc}^{(2)})$
6	`MUL`	`2`	$\mathsf{acc}^{(4)} := \mathsf{acc}^{(3)} * 2$
7	`ADD`	`temp`	$\mathsf{acc}^{(5)} := \mathsf{acc}^{(4)} + \mathit{temp}^{(0)}$
8	`ST`	`temp`	$\mathit{temp}^{(1)} := \mathrm{u8}(\mathsf{acc}^{(5)})$
	⋮		⋮
38	`ADD`	`temp`	$\mathsf{acc}^{(29)} := \mathsf{acc}^{(28)} + \mathit{temp}^{(6)}$
39	`ST`	`from_byte`	$\mathit{from\_byte}^{(0)} := \mathrm{u8}(\mathsf{acc}^{(29)})$

Fig. 1. Program `FromByte` and equivalent SSA form

byte variable `var` a new instance $var^{(j+1)}$ is created assuming that there are already j instances. The symbol x represents either a constant or an existing LHS and is unchanged.

For now, this approach is limited to integer arithmetic and Boolean logic. Converting to other types results in separate invariants for all possible integer values.

For the `FromByte` program the results are presented on the right side of Fig. 1. In the first line, the instance $\mathsf{acc}^{(0)}$ of the accumulator is created, resembling the load instruction of the variable `in0`. After the 39 lines of the program, the expression $\mathrm{u8}(\mathsf{acc}^{(29)})$ (a `BYTE` cast of the 29th instance of the accumulator) is assigned to the output variable `from_byte`. Based on this final expression, we will derive the invariant of this program in the next section.

3 Generation of Invariants

In this section we will use the SSA form defined in the last section to generate symbolic invariants. The key idea is that for each variable introduced in SSA, we still have symbolic information how the value was calculated if we inspect the corresponding RHS. By inducing and thereby replacing all LHSs with the corresponding RHS expressions until we reach constants or variables with an instance number of 0, we can build up symbolic expressions for each LHS. If we do this for the output variable `from_byte` of the example program at the end of the PLC cycle, we obtain the following resolution steps:

`LD x`	$\mathsf{acc}^{(i+1)} := x$
`ST var`	$var^{(j+1)} := \mathrm{u8}(\mathsf{acc}^{(i)})$
`ADD x`	$\mathsf{acc}^{(i+1)} := \mathsf{acc}^{(i)} + x$
`SUB x`	$\mathsf{acc}^{(i+1)} := \mathsf{acc}^{(i)} - x$
`AND x`	$\mathsf{acc}^{(i+1)} := \mathsf{acc}^{(i)}$ and x

Fig. 2. IL instructions and how the are represented in SSA form

$$
\begin{aligned}
\texttt{from\_byte} &= \mathit{from\_byte}^{(0)} \\
&= \mathrm{u8}(\mathsf{acc}^{(29)}) \\
&= \mathrm{u8}(\mathsf{acc}^{(28)} + \mathit{temp}^{(6)}) \\
&= \mathrm{u8}((\mathsf{acc}^{(27)} * 128) + \mathsf{acc}^{(25)}) \\
&= \mathrm{u8}((\texttt{in7} * 128) + \mathrm{u8}(\mathsf{acc}^{(24)} + \mathit{temp}^{(5)})) \\
&\ldots \\
&= \mathrm{u8}((\texttt{in7} * 128) + \mathrm{u8}((\texttt{in6} * 64) + \mathrm{u8}((\texttt{in5} * 32) + \mathrm{u8}((\texttt{in4} * 16) \\
&\quad + \mathrm{u8}((\texttt{in3} * 8) + \mathrm{u8}((\texttt{in2} * 4) + \mathrm{u8}((\texttt{in1} * 2) + \mathrm{u8}(\texttt{in0}))))))).
\end{aligned}
$$

Here, each step consists of a replacement of a LHS by its corresponding RHS expression by a look-up in the list of SSA expressions. In the step from the second to the third line, e. g., $\mathsf{acc}^{(29)}$ was replaced by $\mathsf{acc}^{(28)} + \mathit{temp}^{(6)}$ (cf. Fig. 1 line 38). Afterwards, we can apply some simplifications on the expression found, to make them more readable and easier to understand.

This includes

- the folding of constants (e. g. (*false* or *false*) is rewritten as *false*),
- the removal of unnecessary casts,
- the elimination unused subexpressions (e. g. (*true* or *expr*) is rewritten as *true*).

These steps are repeated until there are no further simplifications.

The crucial point here is that the final invariant for `form_byte` is exactly the invariant that was manually specified as the LTL formula (1) by Pavlovic et al. This means we can derive the invariant without a time-consuming model checking process here. Often, these invariants also give insight into the program behavior without manual writing specifications.

Now, we will formalize the derivation of the invariants. For each non-temporary program variable `var`, we have some instance $var^{(0)}$ which corresponds to the value of the variable at the beginning of the PLC cycle. This can be either an input value read from a sensor or the last value of the previous cycle. For each output variable $\texttt{out}_i$, on the other hand, we have some final value ${out_i}^{(n_i)}$, where n_i is maximal. This corresponds to the final output value at the end of the PLC cycle. If we derive the invariants for all variables ${out_i}^{(n_i)}$ of the program using this technique, we get the dependence — for

```
VAR_INPUT  X: BYTE; END_VAR
VAR        Y: BYTE; END_VAR
      LD     X
      GT     100
      JMPC   m
      LD     Y
      ADD    1
      ST     Y
      RET
m:    LD     Y
      SUB    1
      ST     Y
      RET
```

Fig. 3. Example program with input-dependent invariants

each cycle — of the new variable values on the old values (possibly input values) as symbolic invariants.

These invariants sometimes depend on the actual values of the inputs, which we will now show on an additional example. The program shown in Fig. 3 has two variables X, Y of type BYTE, where X is an input variable and Y is an internal variable. In each cycle, Y is decremented if X is greater than 100, otherwise incremented. Since this program has two different execution paths for $\texttt{X} \in [0, 100]$ and $\texttt{X} \in [101, 255]$, we get two different invariants for this program. Generating the invariants for the second program, results in the two invariants

$$\begin{aligned} \texttt{X} \in [0, 100] &\implies Y^{(1)} = \text{u8}(Y^{(0)} + 1) \\ \text{and } \texttt{X} \in [101, 255] &\implies Y^{(1)} = \text{u8}(Y^{(0)} - 1) \end{aligned}$$

for the distinct execution paths, where $Y^{(0)}$ and $Y^{(1)}$ are the values in Y before and after executing the cycle.

As a real-world example, we also generated invariants for an implementation of the PLCopen safety function block *emergency stop* [16], kindly provided by Soliman and Frey [18]. The function block has 5 Boolean inputs and the implementation uses 11 internal variables (giving 2^{16} possible configurations). The implementation uses the internal variables to control the current state while monitoring the emergency stop signal, the reset function, etc. With our method, we derive 75 invariants for the function block. A typical invariant is:

$$\begin{array}{rrcl} & \mathit{Activate}^{(0)} = \mathit{false} & & \mathit{Ready}^{(1)} = \mathit{false} \\ \wedge & \mathit{S_1}^{(0)} = \mathit{false} & \implies & \wedge\ \mathit{S\_EStopOut}^{(1)} = \mathit{false} \\ \wedge & \mathit{S_2}^{(0)} = \mathit{true} & & \wedge\ \mathit{S_1}^{(1)} = \mathit{true} \wedge \mathit{S_2}^{(1)} = \mathit{false}. \end{array}$$

The variables S_1 and S_2 indicate the states *idle* and *init*, while the other variables are the input and output variables according to the PLCopen specification. The variables of the program not shown are not relevant for the invariant. By this invariant, we can deduce that if we are in state *init* and the `Activate` signal is reset, the outputs `Ready` and `S_EStopOut` are reset and the state *idle* is signaled, independently of all other inputs or program states.

To gain an overview over program behavior, we provide a means for filtering the invariants for certain inputs or outputs. There are, for instance, only 10 invariants generated where the input `Activate` is *false*. Similar invariants can be inspected for the state variables `S_i` or output variables like `S_EStopOut`.

4 Related Work

Transferring formal verification methods from theory to practical applications is an active topic, and particularly important for safety-critical systems. This includes, but is not limited to model checking [17] and static program analysis [3,8]. Additionally, techniques based on exact decision procedures have found application, for instance, the work by Sülflow and Drechsler [19] on SAT-based equivalence checking. More comprehensive overviews of existing approaches for formlization and verification of PLC programs are given elsewhere [2,13,1].

Invariant generation using program rewriting is a widely appreciated concept in the fundamental research on program verification [14], e.g., in the context of verifying heap-manipulating programs [11]. However, to the best of our knowledge, the concept of applying rewriting logic was not applied to PLC programs before, where it allows to derive strong invariants due to the limitations of the underlying hardware platform. Often, even simple logics as the one described in this work suffice for deriving expressive invariants.

5 Discussion & Future Work

Invariants are a simple means to represent properties of programs. The logic considered in this paper — in contrast to other logics used for specifying properties such as LTL or CTL — is easy to formulate and understand. Usually, invariants are used in formal verification techniques like model checking to specify properties to be proven. In this paper, we proposed a new approach that automatically derives invariants for all variables of a program, i. e., an over-approximation of properties of the variables is generated automatically. Thus, users do not need to provide properties of a program and use formal verification techniques to prove these properties, which might be error-prone and lengthy. The invariants derived can then be analyzed by users to check whether the program behaves as expected.

Two specifics enabled the automatic generation of invariants for PLC programs and facilitated the scalability of the approach described: the underlying hardware of the PLCs and the usage of SSA form. The underlying hardware of PLCs is simple due to its use in safety-critical systems, well structured, and extensively documented. The use of a SSA form enables to come up with invariants, which are, for example, only valid after the last iteration of a loop during a PLC cycle as in the case of `from_byte` variable. This is especially helpful in PLC programs as the values of variables are only visible after the execution of a complete cycle and not during the cycle.

As shown by our examples, the approach delivers invariants that capture the behavior of programs in a way easily accessible by users. Additionally, the examples show that the approach also scales well for larger programs.

There are several directions for future improvement. First, not all properties can be expressed using invariants and it is therefore useful to investigate whether the same approach can be used to automatically generate properties using a more complex logic. Furthermore, users want to check whether their properties are satisfied by the program. This can be achieved by checking whether their invariants are entailed by the invariants generated by our approach. If the user-provided invariants are satisfied, it could be useful to also provide information about the additional properties which are satisfied by the program but not specified by the users. This process could be supported by a graphical representation of the invariants.

References

1. M. Bani Younis and G. Frey. Formalization of existing PLC programs: A survey. In *CESA*, 2003.
2. L. Baresi, M. Mauri, A. Monti, and M. Pezze. PLCTools: Design, formal validation, and code generation for programmable logic controllers. In *SMC*, pages 2437–2442, 2000.
3. S. Bornot, R. Huuck, B. Lukoschus, and Y. Lakhnech. Utilizing static analysis for programmable logic controllers. In *ADPM*, pages 183–187, 2000.
4. A. Cimatti, E. M. Clarke, E. Giunchiglia, F. Giunchiglia, M. Pistore, M. Roveri, R. Sebastiani, and A. Tacchella. NuSMV version 2: An opensource tool for symbolic model checking. In *CAV (2002)*, volume 2404 of *LNCS*, pages 241–268. Springer, 2002.
5. R. Cytron, J. Ferrante, B. K. Rosen, M. N. Wegman, and F. K. Zadeck. Effciently computing static single assignment form and the control dependence graph. *ACM Trans. Program. Lang. Syst.*, pages 451–590, 1991.
6. E. A. Emerson. *Handbook of Theoretical Computer Science*, volume B, chapter Temporal and Modal Logics, pages 995–1072. The MIT Press, 1991.
7. C. A. R. Hoare. An axiomatic basis for computer programming. *Comm. of the ACM*, 12(10):576–585.
8. R. Huuck. *Software Verification for Programmable Logic Controllers.* Dissertation, University of Kiel, Germany, April 2003.
9. International Electrotechnical Commission. *IEC 61131-3 Ed. 1.0: Programmable Controllers — Part 3: Programming languages.* International Electrotechnical Commission, Geneva, Switzerland, 1993.
10. International Electrotechnical Commission. *IEC 61508: Functional Safety of Electrical, Electronic and Programmable Electronic Safety-Related Systems.* International Electrotechnical Commission, Geneva, Switzerland, 1998.

11. S. K. Lahiri and S. Qadeer. Back to the future: revisiting precise program verification using SMT solvers. In *POPL*, pages 171–182. ACM, 2008.
12. T. Mertke and G. Frey. Formal verification of PLC-programs generated from signal interpreted petri nets. In *2001 IEEE International Conference on Systems, Man, and Cybernetics*, volume 4, pages 2700–2705. IEEE Computer Society Press, 2001.
13. T. Mertke and T. Menzel. Methods and tools to the verification safety-related control software. In *SMC*, pages 2455–2457, 2000.
14. G. Nelson. Verifying reachability invariants of linked structures. In *POPL*, pages 83–47. ACM, 1983.
15. O. Pavlovic, R. Pinger, and M. Kollmann. Automated formal verification of PLC programms written in IL. In *VERIFY*, number 259 in Workshop Proce., pages 152–163. CEUR-WS.org, 2007.
16. PLCopen TC5. *Safety Software Technical Specification, Version 1.0, Part 1: Concepts and Function Blocks.* PLCopen, Germany, 2006.
17. B. Schlich, J. Brauer, J. Wernerus, and S. Kowalewski. Direct model checking of PLC programs in IL. In *DCDS*, 2009. To appear.
18. D. Soliman and G. Frey. Verification and validation of safety applications based on PLCopen Safety Function Blocks using Timed Automata in UPPAAL. In *DCDS*, 2009. To appear.
19. A. Sülflow and R. Drechsler. Verification of plc programs using formal proof techniques. In *FORMS/FORMAT*, pages 43–50. L'Harmattan, 2008.

Automatic Fault Localization for Programmable Logic Controllers *

André Sülflow and Rolf Drechsler

Institute of Computer Science, University of Bremen
28359 Bremen, Germany
{suelflow, drechsle}@informatik.uni-bremen.de

Abstract. *Programmable Logic Controllers* (PLCs) are widely applied to control safety critical systems. Efficient formal and non-formal methods to detect faulty behavior have been developed, but finding the cause of the buggy behavior is often still a manual process. Automatic fault localization for PLCs is studied in this paper. Methods for automated debugging are analyzed and compared with respect to accuracy and run time. The experimental results on industrial models show a high accuracy at low run time costs.

Keywords: Debugging, Boolean SAT, Program slicing, Programmable Logic Controllers

1 Introduction

A *Programmable Logic Controller* (PLC) is a re-programmable computer based on sensors and actors that is running a user defined software periodically. That makes a PLC highly configurable and applicable in various industrial sectors, e.g., in nuclear power plants and in railway interlocking systems. The assurance of the correct behavior in safety critical systems is a must. In this work PLCs suitable to control railway electronic interlocking specified to *Safety Integrity Level 3* (IEC61508) are considered.

Model checking of PLC software was proposed in, e.g., [3,11,15,18]. The output of a model checker is either a proof of correctness of the model with respect to a specification or a failure trace, i.e., a counter-example that shows the incorrect behavior. Debugging the observed faulty behavior often relies on manual simulation and is a time consuming task. Automated debugging of faulty behavior in PLC programs has not been considered so far and is in focus of this paper.

Several techniques for automated debugging have been proposed for software (e.g. [22,2,13,9]) as well as for hardware (e.g. [21,6,17,7]). The aim of

* This work has been supported in part by the Rail Automation Graduate School (Siemens Transportation, Braunschweig, Germany) and the European Union (project DIAMOND, FP7-2009-IST-4-248613).

E. Schnieder, G. Tarnai (eds.), *FORMS/FORMAT 2010*,
DOI 10.1007/978-3-642-14261-1_25, © Springer-Verlag Berlin Heidelberg 2011

automated debugging is the highlighting of potential fault locations to an engineer to reduce the complexity for a subsequent manual debugging session. Thus, fixing the faulty behavior remains a manual task to avoid unexpected changes that may be introduced by methods that perform repairs automatically [4].

Explaining debugging techniques compute traces that are similar to a failure trace but fulfill the specification to explain the faulty behavior [10]. The difference between the execution of the correct trace and a failure trace reveals potential fault candidates. Program slicing returns statements on the path from a (faulty) observation point to the primary inputs. The analysis is performed statically or dynamically with respect to the actual values of the failure trace [22,2]. The techniques above do not fully exploit the expected behavior at the observation points and an over-approximation of fault candidates may be returned only.

Model-based diagnosis is more precise by computing fault locations on an abstract model [13]. Using a solver for *Boolean Satisfiability* (SAT) allows to handle large problem instances due to the tremendous improvements in Boolean satisfiability solving [5,12]. SAT-based debugging [17] partially automates the debugging by finding possible fault locations, i.e., components that can *fix* the faulty behavior. The usability for complex models was shown for debugging hardware [17] as well as software [9].

In this work we analyze methods for automated debugging of PLC programs. Three automated debugging methods are evaluated: (1) static analysis, (2) dynamic analysis, and (3) correction-based debugging.

2 Diagnosis Model

In IEC61131-3 two textual and three graphical programming languages for PLCs are standardized. In this paper we focus on the assembler-like language *Instruction List* (IL). More specifically, *Statement List* (STL) is the input language of the PLC considered [16]. STL is similar to IL and is referenced as IL in the remainder of this paper.

The behavior of an IL program is specified in M lines of code that are sequentially executed in a deterministic order. Each line contains one instruction, composed of an optional label, an operator and an optional operand of $\{variable, constant\}$. The control flow is influenced by jump instructions, that are referencing a label in line r, $1 \leq r \leq M$. IL programs may be additionally divided into sub-programs.

A CDFG is constructed from the IL program to perform automated debugging. Nodes in the CDFG represent data instructions or control predicates, respectively. Data dependency and the control flow are modeled by edges.

An abstract model or a concrete model of the executing CPU of the PLC can be used to construct the CDFG. Because a CPU consists of several registers, stacks as well as accumulators the execution of an IL instruction highly depends on those state variables. The usage of a concrete model for the CPU semantic is more complex, but allows to perform CPU specific analyses. For example, registers may be highlighted to observe the faulty behavior, e.g., for a subsequent manual debugging session. A concrete model of the SIMATIC S7 CPU is used in this work. The SIMATIC S7 operates on a 16-bit status word, two 32-bit accumulators, and a nesting stack that stores intermediate results. Operations up to a bit-width of 32 are supported [16].

Technically, the IL program is translated to *SystemC* [20] by augmenting the behavioral information of the underlying CPU as described in [18]. Afterwards, the augmented implementation is analyzed with the parser of [8] to construct a CDFG representation similar to a netlist on RTL [19]. Depending on the chosen instruction, one instruction in IL corresponds to $t, 0 < t$, nodes in the CDFG. For example, the instruction L (Load) operates on the accumulator and requires two assign nodes (one for each accumulator). Other instructions operate on the 16-bit status word of the CPU, e.g., JC (Jump Conditional) evaluates and updates four bits of the status word. The reference to the original instruction in IL is kept for each node in the CDFG. Thus, the selection of at least one of the t nodes of an instructions enables to mark the instruction itself.

Faults in software that change the output behavior of the IL program and that are observable at least at one observation point are considered. Thereby, an observation point may be any program state, internal variable, or primary output. Methods for verification are capable to provide one failure trace or in more general a set of m failure traces. The cause of faults in PLC hardware is most likely physical (e.g. due to aging) than logical (e.g. a missing instruction). However, the extension to debug faults in the underlying hardware is a possible extension for future work.

Without loss of generality, let a failure trace consists of:

- an input stimuli I to activate the fault,
- a set of R observation points OP_i and its faulty responses under I: $v_{faulty}(OP_i)$, $1 \leq i \leq R$,
- a set of R expected responses for each observation point: $v_{expected}(OP_i)$, $1 \leq i \leq R$

An input stimuli I defines values for primary inputs and values for state variables with respect to a single PLC cycle. The faulty behavior is observable at least at the R observation points. Thus, simulating the PLC program with respect to the input stimuli I yields a pairwise distinct at all observation points: $\forall i : v_{faulty}(OP_i) \neq v_{expected}(OP_i), 1 \leq i \leq R$. The expected responses are automatically obtained by simulating the stimuli I on a reference model.

Components are used to explain the faulty behavior. In general, a component may be of any granularity, e.g., a set of instructions, a single instruction, or a single operand. However, the complexity of diagnosis increases with the granularity. Without loss of generality, single instructions in IL are considered as components in this work only.

A *fault candidate* is a component that *may change* the value at all observation points to the expected values. A *fix* is a component that *changes* the value at all observation points to the expected values: $\forall i : v_{faulty}(OP_i) = v_{expected}(OP_i), 1 \leq i \leq R$. The actual fault site is the position where the fault was injected and it is called *fault site*.

3 Debugging Algorithms

The CDFG and a set of m failure traces ($m \geq 1$) are the input for automatic debugging. Three debugging techniques are analyzed in the following: (1) static analysis, (2) dynamic analysis, and (3) correction-based debugging.

The following sections briefly introduce the methods. For more information about the debugging techniques, we refer the reader to the referenced papers in the respective sections.

3.1 Static Analysis

Independent of any failure trace, a static analysis on the CDFG enables to identify components that have no influence on the output behavior of the IL program. The redundant code fragments manifest themselves as pending nodes in the CDFG, i.e., nodes without successors, and are detectable in linear time (linear in the number of nodes). Only nodes that are in at least one fan-in of any primary output influence the output behavior. Operations that are not in any fan-in are redundant, cannot change the output behavior, and can be removed from consideration. By this, the initial number of fault candidates is reduced. Therefore, the information is also worthwhile for code optimization on the IL program. For example, an assignment to a variable A is removable, if no read operation is applied to A and A is not an output or a state element.

An additional analysis based on static slicing [22] uses the observation points to further reduce the number of fault candidates. A faulty behavior is observable at an observation point OP_i. Thus, all nodes on the path to any OP_i are potential fault candidates. All nodes that are not in the recursive fan-in of any OP_i cannot influence the value at OP_i and do not have to be considered for diagnosis. The nodes on the path from an observation point OP_i are recursively computed in linear time.

Let P_{ji} be a set of nodes on the path for failure trace j and observation point OP_i. While assuming a single fault, all relevant nodes for debugging (P') are computed by $P' = \bigcap P_{ji}$. For multiple faults the intersection can be empty, e.g., if a faulty behavior is observed on two observation points that have disjunctive input cones. One element from each path has to be selected while assuming multiple faults. However, due to the computational complexity for computing the hitting set, an over-approximation is considered for multiple faults only: $P' = \bigcup P_{ji}$. The union of all nodes from any path are returned as fault candidates in case of multiple faults.

3.2 Dynamic Analysis

Dynamic analysis is applied to debug the IL program with respect to the input stimuli of a failure trace. Instead of debugging all parts of an IL program, the failure trace is analyzed to determine the *sensitive path.*

For this purpose, the failure trace is simulated on the CDFG to obtain the assigned values on each node in the CDFG at first. Afterwards, an affect analysis is performed based on path tracing [1], i.e., a path from an observed faulty behavior at an observation point to the primary inputs is computed [21]. The path is extracted by following the controlling values, i.e., the operations that are responsible for the current value at the observation point $v_{faulty}(OP_i)$. In difference to dynamic slicing [2], path tracing returns statements that are *responsible* for the current value at the observation points instead of statements that *might affect* the value only.

For example, the controlling paths for a logical *AND* with output value 0 are the inputs with value 0 assigned. An input with 1 assigned is not responsible for the output value 0 and thus cannot be responsible for the faulty behavior. If both inputs are controlling paths, the algorithm follows both paths.

The dynamic analysis is performed for all observation points OP_i and all m failure traces in linear time. Thereby, each failure trace is simulated on the CDFG, followed by path tracing. All nodes on the sensitive path from any observation point are potential fault candidates. Like for the static analysis, either the intersection or the union is computed for all sensitive paths while assuming single faults or multiple faults, respectively [21].

3.3 Correction-Based Debugging

Static analysis and the dynamic debugging technique use information about the observation points OP_i and the input stimuli only. Correction-based debugging is a technique to obtain higher accuracy by considering additionally the expected output responses at the observation points OP_i [17]. The input for the debugging algorithm is a CDFG, a set of m failure traces and the expected responses for each observation point $v_{expected}(OP_i)$.

In contrast to static analysis and debugging based on simulation, the returned fault candidates are proved to fix the failure trace while assuming a non-deterministic behavior of a component. That is, the output behavior of a component is replaced by an unconstrained new input that may have any value assigned. This allows to check whether the application of any correction at the component fixes the faulty behavior at all observation points, i.e., it holds $\forall i : v_{expected}(OP_i) = v_{faulty}(OP_i), 1 \leq i \leq R$.

For each component *correction logic* is inserted in the CDFG as described in [9]. That is, each right-hand side of an expression and all control predicates can be replaced by a non-deterministic value. One *abnormal predicate* for each component controls the activation at all expressions and control predicates

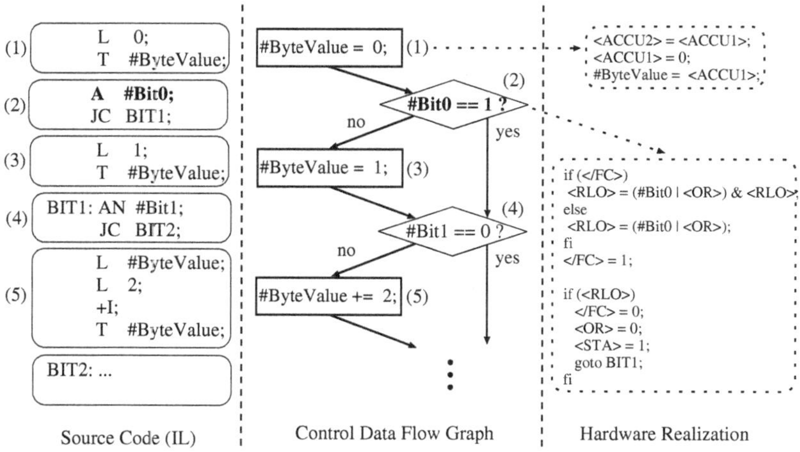

Fig. 1. Faulty program

from the same component, simultaneously. The CDFG is extended by correction logic, the input stimuli (I) is constrained to the values of the failure trace, and the observation points are restricted to the expected responses ($v_{expected}(OP_i)$).

The initial debugging instance is contradictory, due to the constrained expected responses at the observation points OP_i. Correction-based debugging resolves the conflict by activating exactly k abnormal predicates. Thereby, k gives the cardinality of the fault candidate and may be a single fault ($k = 1$) or a multiple fault ($k > 1$). Starting with $k = 1$, k is incremented as long as the instance is still contradictory. A *cardinality constraint* controls the activation of abnormal predicates. The debugging model is translated into *Conjunctive Normal Form* (CNF) and given to a *Boolean satisfiability solver* (SAT solver) to declare the instance satisfiable or unsatisfiable. After the conflict is resolved, i.e. the instance is satisfiable, fault candidates of minimal cardinality are obtained.

4 Evaluation

The debugging methods are evaluated on concrete benchmarks in the following sections. A detailed case study is presented in Section 4.1 to highlight the pro and cons of the debugging techniques. Experimental results for the application on industrial benchmarks are given in Section 4.2.

4.1 Case Study

An example is used to illustrate the approaches. Figure 1 (left) shows a part of a combinational IL program. The program converts the eight inputs ($Bit0$

Table 1. Diagnosis results

Test case		Static Analysis						Dynamic Analysis						Correction-Based					
Bit0	Bit1	(1)	(2)	(3)	(4)	(5)	LOC	(1)	(2)	(3)	(4)	(5)	LOC	(1)	(2)	(3)	(4)	(5)	LOC
0	0	x	x	x	x	x	12		x	x	x		6		x	x			4
1	0	x	x	x	x	x	12	x	x		x		6	x	x				4
0	1	x	x	x	x	x	12		x	x	x	x	10		x	x		x	8
1	1	x	x	x	x	x	12	x	x		x	x	10	x	x			x	8
Single fault		x	x	x	x	x	12		x		x		4		x				2
Multiple fault		x	x	x	x	x	12	x	x	x	x	x	12		x				2

to *Bit*7) into a byte (*ByteValue*). The inputs (*Bit*0 to *Bit*7) are Booleans and the output (*ByteValue*) is a byte of bit-width 8. The program checks each input bit separately and adds the corresponding value to *ByteValue*.

The program is faulty at checking the value of *Bit*0 (Line 3, marked bold). Here, the logical AND (*A*) is used instead of the (correct) NAND operator (*AN*). Therefore, the least significant bit (*Bit*0) is not interpreted correctly and the condition in code block (2) inverts. The computed *ByteValue* is either too high or too low with a delta of one.

To demonstrate the techniques, the program is partially defined and requires input values for *Bit*0 and *Bit*1, only. The primary output *ByteValue* is the observation point in the example. Simulating the test case $Bit0 = 1$ and $Bit1 = 0$ leads to the faulty value $v_{faulty}(\text{ByteValue}) = 0$ whereas $v_{expected}(\text{ByteValue}) = 1$ is expected.

Figure 1 (center) shows an abstract CDFG for the IL program. The numbers in brackets are referencing the corresponding code block in IL (Figure 1 (left)). For example, the two instructions in code block (1) initialize *ByteValue* with 0.

The details of the register changes of the CPU are exemplarily shown for the instructions in block (1) and block (2) on the right-hand side of Figure 1. The internal registers of the CPU are enclosed in '<' and '>'. For example, the two accumulators are denoted by '<ACCU1>' and '<ACCU2>'. As shown in Figure 1, the operations use and modify different registers. Thus, a fault candidate gives additional information to the internal registers where the fault is observable best. Moreover, a fine granular component model may further increase the accuracy of diagnosis by highlighting value changes on register level as potential fault candidates.

Table 1 compares the quality of the debugging results with respect to all possible test cases ($2^2 = 4$) (Column *Test case*). The rest of the columns give the diagnosed code blocks ((1), .., (5)) for the algorithms and the corresponding number of lines in the IL program (Column *LOC*). An 'x' marks the diagnosis with respect to the test case. The final diagnosis results for the consideration of all four test cases with respect to a single fault assumption and a multiple fault assumption is given in the last two rows. Note, code blocks are highlighted, but the diagnosis is itself performed on instruction level. An (un)marked code block means that none/all instructions within the code block have been marked as fault candidate by the debugging algorithm. None of the debugging algorithms return a subset of the instructions within a code block only.

The application of static analysis on the example is not very efficient. That is, any path from the primary inputs ends at the observation point *ByteValue*. All nodes are required for the computation of *ByteValue* and no node is prunable. As shown in Table 1, all diagnoses return all instructions in the IL program as fault candidates.

Dynamic diagnosis is more accurate. The usage of the input stimuli increases the accuracy and some code blocks are pruned (see Table 1). Each test case activates different paths and the final diagnosis depends on the considered test cases. The final diagnosis with respect to a single fault assumption for test case 1 and test case 3 returns code block (2), (3), and (4). The consideration of all four test cases further prunes code blocks and returns code block (2) (the fault site) and code block (4) as potential fault candidates only. However, the cardinality of the fault is typically not known in advance and using all test cases is often not feasible for models with a large number of inputs. Additionally, any fix at (4) cannot fix the faulty behavior. The fault candidates in code block (4) have not been proved to fix the faulty behavior. The final diagnosis without any assumption on the fault cardinality returns all code blocks as potential fault candidates and does not help for debugging.

Correction-based debugging is more accurate in comparison to the first two analyses. Only fault candidates that fix the faulty behavior are returned. Code block (4) is accurately determined to be not a fault candidate. The results for each failure trace are more accurate and the final diagnosis with respect to all test cases returns the block with the original fault site only. The fault candidates are automatically of minimal cardinality and no assumption on the cardinality of the fault has to be made.

4.2 Industrial Software

The debugging algorithms are further evaluated on six industrial programs from the railway interlocking domain. The models have different complexity and use Boolean operations (e.g. logical AND and OR), arithmetic operations (e.g. 16-bit addition), and control flow statements (e.g. conditional jumps). The number of instructions in the models are ranging from 22 to 833 and the models have up to 23 primary inputs, 10 primary outputs, and 31 state variables.

A faulty implementation has been created by injecting a single fault manually in the IL model, e.g. replacing an operator *A* with an *AN*. The failure traces are obtained by applying SAT-based equivalence checking to a high level specification (see [18]). In all cases, the diagnosis is performed with respect to a single fault assumption and a randomly generated single failure trace. Components on instructions level are considered.

Table 2. Efficiency

model	Initial	Static Analysis			Dynamic Analysis			Correction-Based		
	|FCs|	|FCs|	Red.	Time	|FCs|	Red.	Time	|FCs|	Red.	Time
model-1	63	51	19.05	< 1.00	37	41.27	< 1.00	23	63.49	< 1.00
model-2	53	47	11.32	< 1.00	30	43.40	< 1.00	17	67.92	< 1.00
model-3	22	18	18.18	< 1.00	12	45.45	< 1.00	11	50.00	< 1.00
model-4	367	74	79.84	< 1.00	43	88.32	12.81	30	91.83	14.21
model-5	615	130	78.86	1.43	92	85.04	26.52	3	99.51	2.96
model-6	833	728	12.61	9.52	374	55.10	51.99	29	96.52	39.97

All experiments are conducted on an AMD Athlon 6000+ (3GHz) running Linux. The main memory was limited to 4 GB and the run time is measured in CPU seconds. The word level framework WoLFram is the back-end for the analysis [19]. ZChaff is the underlying SAT engine for correction-based debugging [14].

The experimental results are shown in Table 2. Column *|FCs|* gives the number of fault candidates and Column *Red.* denotes the reduction to the initial number of fault candidates in percent. The diagnosis time for the debugging algorithms is shown in Column *Time*.

All debugging algorithms are capable to reduce the initial number of fault candidates. For model-4 and model-5 more than 78% of the instructions are determined by all algorithms to not be a fault candidate. For other benchmarks, static analysis and dynamic analysis provide a rough diagnosis only, e.g. for model-1 and model-2. Here, simulation has advantages over static analysis, but correction-based debugging outperforms both algorithms. Correction-based debugging shows the best diagnostic resolution for all benchmarks. Thereby, the original fault site has been correctly determined by all algorithms. Fault candidates are efficiently detected by all methods and the complexity for a subsequent manual fix of an engineer is reduced. Regarding run time, all debugging algorithms require less than one minute for the diagnosis only.

In summary, all debugging algorithms support the manual debugging task by reducing the number of fault candidates. An engineer can focus on small parts of the whole program only. Automatic debugging gives hints for correction and helps to understand the faulty behavior. The computational overhead is low and the accuracy high.

References

1. M. Abramovici, P.R. Menon, and D.T. Miller. Critical path tracing - an alternative to fault simulation. In *Design Automation Conf.*, pages 214–220, 1983.
2. H. Agrawal and J.R. Horgan. Dynamic program slicing. In *PLDI '90: Proceedings of the ACM SIGPLAN 1990 conference on Programming language design and implementation*, pages 246–256, 1990.
3. G. Canet, S. Couffin, J.-J. Lesage, A. Petit, and P. Schnoebelen. Towards the automatic verificication of PLC programs written in instruction list. In *IEEE conf. on Systems, Man and Cybernetics (SMC)*, pages 2449–2454, 2000.

4. K.-H. Chang, I.L. Markov, and V. Bertacco. Fixing design errors with counterexamples and resynthesis. *IEEE Trans. on CAD*, 27(1):184–188, 2008.
5. M. Davis, G. Logeman, and D. Loveland. A machine program for theorem proving. *Comm. of the ACM*, 5:394–397, 1962.
6. M. Fahim Ali, A. Veneris, S. Safarpour, R. Drechsler, A. Smith, and M.S.Abadir. Debugging sequential circuits using Boolean satisfiability. In *Int'l Conf. on CAD*, pages 204–209, 2004.
7. G. Fey, S. Staber, R. Bloem, and R. Drechsler. Automatic fault localization for property checking. *IEEE Trans. on CAD*, 27(6):1138–1149, 2008.
8. C. Genz and R. Drechsler. System exploration of SystemC designs. In *IEEE Annual Symposium on VLSI*, pages 335–340, Mar. 2006.
9. A. Griesmayer, S. Staber, and R. Bloem. Automated fault localization for C programs. *Electronic Notes in Theoretical Computer Science*, 174(4):95–111, 2007.
10. Alex Groce, Sagar Chaki, Daniel Kroening, and Ofer Strichman. Error explanation with distance metrics. *Int. J. Softw. Tools Technol. Transf.*, 8(3):229–247, 2006.
11. R. Huuck. *Software Verification for Programmable Logic Controllers.* PhD thesis, Faculty of engineering, University of Kiel, Germany, 2003.
12. J.P. Marques-Silva and K.A. Sakallah. GRASP: A search algorithm for propositional satisfiability. *IEEE Trans. on Comp.*, 48(5):506–521, 1999.
13. W. Mayer and M. Stumptner. Model-based debugging - state of the art and future challenges. *Electronic Notes in Theoretical Computer Science*, 174(4):61–82, 2007.
14. M.W. Moskewicz, C.F. Madigan, Y. Zhao, L. Zhang, and S. Malik. Chaff: Engineering an efficient SAT solver. In *Design Automation Conf.*, pages 530–535, 2001.
15. O. Pavlovic, R. Pinger, M. Kollmann, and H.-D. Ehrich. Principles of formal verification of interlocking software. In *Symposium on Formal Methods for Automation and Safety in Railway and Automotive Systems (FORMS/FORMAT)*, pages 370–378, 2007.
16. Siemens. *SIMATIC–Statement List (STL) S7-300 and S7-400 Programming*, 2003.
17. A. Smith, A. Veneris, M. Fahim Ali, and A.Viglas. Fault diagnosis and logic debugging using boolean satisfiability. *IEEE Trans. on CAD*, 24(10):1606–1621, 2005.
18. A. Sülflow and R. Drechsler. Verification of PLC programs using formal proof techniques. In *Symposium on Formal Methods for Automation and Safety in Railway and Automotive Systems (FORMS/FORMAT)*, pages 43–50, 2008.
19. A. Sülflow, U. Kühne, G. Fey, D. Große, and R. Drechsler. WoLFram - a word level framework for formal verification. In *International Symposium on Rapid System Prototyping (RSP)*, pages 11–17, 2009.
20. Synopsys Inc. and CoWare Inc. and Frontier Design Inc. *Open SystemC Initiative.* http://www.systemc.org, 2008.
21. A. Veneris and I. N. Hajj. Design error diagnosis and correction via test vector simulation. *IEEE Trans. on CAD*, 18(12):1803–1816, 1999.
22. M. Weiser. Program slicing. *IEEE Trans. Software Engineering*, 10(4):352–357, 1984.

Author Index

E. Schnieder, G. Tarnai (eds.), *FORMS/FORMAT 2010*,
DOI 10.1007/978-3-642-14261-1, © Springer-Verlag Berlin Heidelberg 2011

FSC
www.fsc.org
MIX
Papier | Fördert gute Waldnutzung
FSC® C083411

Kolibri 360 GmbH
Ferdinand-Jühlke-Straße 7
99095 Erfurt, Deutschland
produktsicherheit@kolibri360.de